Web Microanalysis of Big Image Data

Peter Bajcsy • Joe Chalfoun • Mylene Simon

Web Microanalysis of Big Image Data

 Springer

Peter Bajcsy
National Institute of Standards
and Technology
Gaithersburg, MD, USA

Joe Chalfoun
National Institute of Standards
and Technology
Gaithersburg, MD, USA

Mylene Simon
National Institute of Standards
and Technology
Gaithersburg, MD, USA

ISBN 978-3-319-87533-0   ISBN 978-3-319-63360-2   (eBook)
https://doi.org/10.1007/978-3-319-63360-2

Printed on acid-free paper

This Springer imprint is published by Springer Nature
The registered company is Springer International Publishing AG
The registered company address is: Gewerbestrasse 11, 6330 Cham, Switzerland

Preface

We motivate big data microscopy experiments and then introduce the theoretical and architectural underpinnings of our Web Image Processing Pipeline (WIPP) system for analyzing images collected during big microscopy experiments. This book comes with both the WIPP tool and test image collections, in order to increase the reader's understanding and gain experience with practical tools for analyzing big image experiments. We will describe (a) WIPP functionalities, (b) use cases, and (c) components of the web software system (web server and client architecture, algorithms, and hardware-software dependencies). Our descriptions of technical details will follow a top-down presentation and will explain the interactions of the web system components and their impact on computational scalability, provenance information gathering, interactive display, and computing.

Our purpose is to encourage graduate students, postdoctoral students, and scientists to perform big data microscopy experiments. We will attempt to achieve this by providing educational materials, software tools, and test data at the intersection of research areas known as microscopy image analyses and computational science. Furthermore, by providing the WIPP software and test data, students and scientists are empowered with tools to make discoveries with much higher statistical significance than before. Once they become familiar with the web image processing components, they can extend and re-purpose the existing software for new types of analyses.

While there have been a multitude of books about microscopy image processing, there is increasing interest in running these processing algorithms on big microscopy image data. However, when analyzing big data microscopy experiments, scientists are restricted by the image processing methods designed for desktop computers, the time it takes to complete desktop intensive processing, and the complexity of the required big data computational infrastructure. We hope that our readers will find this book to be a useful resource when learning about solutions that can overcome these restrictions.

We ordered the chapters so that readers are first introduced to the problem of big data microscopy experiments (Chap. 1), can install open-source software and become familiar with the capabilities of the web image processing pipeline

Table 1 Book features and associated reader benefits

Features of the book	Corresponding benefits
Design choices and trade-offs in web image processing pipeline	Insights relevant to conducting research with very large images
Open-source software for web image processing	A tool available to a reader for processing big data microscopy experiments
Test data	Educational material for hands-on experience
Open-source software for algorithms running in the image processing pipeline	Published, documented, and evaluated algorithms readily available to a reader for big data applications

(Chap. 2), and then learn about several use cases for the image processing pipeline (Chap. 3). Scientists interested in understanding big image analytics can then proceed to a description of the web system architecture (Chap. 4), the image processing algorithms currently provided in WIPP (Chap. 5), and approaches to accelerate the algorithmic execution (Chap. 6).

The first three chapters are meant for users who would just use the WIPP system. Chapters 4, 5, and 6 are intended for readers who are interested in the computer science and information technology (IT) aspects of big image analytics. The information presented in Chaps. 4, 5, and 6 is also useful for developing, extending, and maintaining the WIPP system because the underlying software and hardware technologies are rapidly changing.

Each chapter follows a top-down presentation. We start with a short introduction and a classification of related methods. We then present a description of the specific methods used in the accompanying software. For several topics, we present an example on how the specific method is applied to a dataset (parameters, computer memory requirements, processing efficiency). Some tips and notes are provided as practical suggestions to improve accuracy or computational performance.

Our intended audience is graduate students, postdoctoral students, and scientists whose research can benefit from big data microscopy experiments. This audience can be drawn from the disciplines that use microscopes such as biomedical sciences, materials sciences, and crop sciences. We envision this as a textbook for a short course in the X-informatics majors where X stands for bio, medical, plant, material, or any other domain that uses microscope imaging as an observational method.

This book may also be of interest to scientists in research laboratories and bio- and material-manufacturing companies. Microscopes in research labs and manufacturing environments generate a large quantity of images regardless of the laboratories' discipline-specific focus. For many laboratories, the ability to deploy and operate an internal web image processing pipeline and associated tools, and perhaps customize them for local processing, will be useful. In these situations, the book can serve as a reference to laboratory scientists.

The primary features of this book and its corresponding benefits for readers are summarized in Table 1. The benefits range from theoretical insights to practical experiences. We will primarily focus on examples from cell biology, but the tools and theoretical foundations are applicable to many other fields in which large-scale image processing and analysis are needed as well.

Gaithersburg, MD, USA   Peter Bajcsy
                                                                                        Joe Chalfoun
                                                                                      Mylene Simon

Terminology

Term	Description
Calibration	Provides a pixel-to-real-distance conversion factor
Dynamic cell	Refers to cellular growth, differentiation, and adaptation to changing circumstances
Image geometry correction	An image manipulation such that the image's projection precisely matches a specific projection surface or shape (i.e., equal areas of projection surface are equal areas in the source image).
Pipeline	A series of processes, usually linear, which filter or transform data. Examples of pipelines in the real world include chaining two or more processes together on the command line using the "I" (pipe) symbol.
Reproducibility	The ability to duplicate an experiment or study, either by the same researcher or by someone else working independently.
Statistical sampling	The selection of a subset of individuals from within a population to estimate characteristics of the whole population.
Workflow	A set of processes, usually nonlinear, often human rather than machine, which filter or transform data, often triggering external events. The processes are not assumed to be running concurrently.

Acknowledgments

WIPP has been developed over the course of 6 years with many contributors from the Information Technology Laboratory (ITL) and Material Measurement Laboratory (MML) at National Institute of Standards and Technology (NIST). The contribution ranged from experimental design, imaging, software design, and image analyses to interpretation, web dissemination, and management. The authors are very grateful to everyone who has participated in the computational science in metrology meetings at NIST because the meetings have served as the forum for discussing ideas and for interdisciplinary education. We would also like to acknowledge the outside NIST contributors who have been willing to share their data and bring many interesting metrology problems to our attention.

Table 1 lists alphabetically sorted names of people we would like to acknowledge that participated in efforts leading up to this book and the accompanying software and datasets.

Our special acknowledgment goes to Alden Dima from the Software and Systems Division at NIST for his invaluable inputs on the book manuscript and his thorough reviews of all chapters.

Table 1 Alphabetically sorted list of people who have contributed to some aspects of the presented material in this book

ITL NIST	MML NIST	Outside of NIST
Amelot, Julien	Bhadriraju, Kiran	Bharti, Kapil
Blattner, Tim	Camp, Charlie	Hoeppner, Daniel
Brady, Mary	Cicerone, Marcus	Hotaling, Nathan
Cardone, Antonio	Elliott, John	Kociolek, Marcin
Dessauw, Philippe	Florczyk, Stephen	Kosecka, Jana
Filliben, James	Halter, Michael	Loser, Wolfgang
Gao, Jing	Lee, Young	McKay, Ron
Gerardin, Antoine	Plant, Anne	Parent, Carol
Juba, Derek	Ritchie, Nicholas	Stuelten, Christina
Keyrouz, Walid	Sarkar, Sumona	Szczypinski, Piotr
Lund, Steve	Schaub, Nicolas	Varshney, Amitabh
Majurski, Michael	Scott, John Henry	Weiger, Michael
Manescu, Petre	Scott, Keana	
Ouladi, Mohamed	Simon, Carl	
Padi, Sarala		
Peskin, Adele		
Vandecreme, Antoine		
Yoon, Soweon		

Disclaimer

Commercial products are identified in this document in order to specify the experimental procedure adequately. Such identification is not intended to imply recommendation or endorsement by the National Institute of Standards and Technology (NIST), nor is it intended to imply that the products identified are necessarily the best available for the purpose.

Abbreviations

ACID properties	Atomicity, consistency, isolation, and durability
Adobe PDF	Adobe Portable Document Format
AJAX	Asynchronous JavaScript And XML
ALU	Arithmetic logic unit
API	Application programming interface
ASP	Active Server Pages
CMOS	Complementary Metal–Oxide–Semiconductor
CPU	Central processing unit
CSS	Cascading Style Sheets
DAG	Directed acyclic graph
DAO	Data access object
DBMS	Database management system
DOM	Document Object Model
DZI	Deep Zoom images
FPGA	Field-programmable gate array
FTP	File Transfer Protocol
GPU	Graphics Processing Unit
HAL	Hypertext Application Language
HATEOAS	Hypermedia as the Engine of Application State
HDF	Hierarchical Definition Format
HTTP	Hypertext Transfer Protocol
HTTPS	Hypertext Transfer Protocol Secure
IT	Information technology
JSON	JavaScript Object Notation
JSP	Java Server Pages
LAN	Local area network
LCD	Liquid-crystal display
MAC	Message authentication code
MPI	Message Passing Interface
MVC	Model-view-controller
NAS	Network-attached storage

NFS	Network File System
NoSQL	Not only SQL
NVD SSD	Non-Volatile Dual memory solid-state device
ODM	Object Document Mapper
OME	Open Microscopy Environment
ONC RPC	Open Network Computing Remote Procedure Call
ORM	Object-Relational Mapping
PCI	Peripheral Component Interconnect
PCI-E	Peripheral Component Interconnect Express
PHP	Hypertext Preprocessor
PKI	Public key infrastructure
PNG	Portable Network Graphics
RAM	Random-access memory
REST	Representational state transfer
RPC	Remote procedure call
SPMD	Single process, multiple data parallel applications
SSD	Solid-state drive
SSL	Secure Sockets Layer
TCP	Transmission Control Protocol
TCP/IP	Transmission Control Protocol/Internet Protocol
TIFF	Tagged Image File Format
TLS	Transport Layer Security
UDP	User Datagram Protocol
UI	User interface
URI	Uniform Resource Identifier
URL	Uniform Resource Locator
W3C	World Wide Web Consortium
WDZT	Web Deep Zoom Toolkit
WFE	Web feature extraction
WIPP	Web Image Processing Pipeline
WMS	Workflow management system
WSM	Web statistical modeling
WWW	World Wide Web
XDR	External Data Representation
XML	eXtensible Markup Language
SAN	Storage area network

Contents

Chapter 1
Introduction to Big Data Microscopy Experiments

1.1 Image Processing Pipeline

An image is an array of picture elements (called pixels) arranged in columns and rows. At every column and row, the pixel has one or more values. For example, images acquired by phase contrast microscopes have single-value pixels, while images collected using bright field or Raman spectroscopy microscopes have multiple-value pixels. Microscope images can be acquired over time to form a video or over multiple z-depths to form a 3D volume. These videos and 3D volumes are still images, each represented by an array of pixels arranged in columns, rows, and either time frames or z-stacks. The image acquisition process is referred to as imaging (see Fig. 1.1) and yields images of a specimen of interest (also denoted as sample). The imaging instrument of interest in this book is a microscope, and the series of computational steps applied to acquired images is the web processing pipeline.

Image processing
Image processing refers to algorithms that take images as inputs and return images as outputs.[1] Image processing typically performs mathematical operations on images to eliminate imaging artifacts, enhance image content, integrate multiple images into the same coordinate system, prepare images for information extraction, or any combination of these operations. For instance, flat field correction eliminates imaging artifacts due to illumination inhomogeneities introduced during imaging. Gaussian filtering enhances image regions buried in Gaussian noise added during imaging by microscope digital circuitry. Segmentation extracts the locations of objects of interest contained in a single image (frequently denoted as one field of view or FOV in microscopy). Image stitching integrates multiple images with a partial spatial overlap to create one large image containing objects of interest spanning a large spatial area.

---

[1] http://www.coe.utah.edu/~cs4640/slides/Lecture0.pdf

© Springer International Publishing AG 2018
P. Bajcsy et al., *Web Microanalysis of Big Image Data*,
https://doi.org/10.1007/978-3-319-63360-2_1

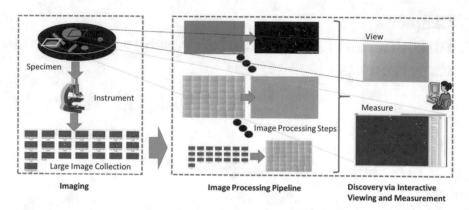

Fig. 1.1 An illustration of web image processing pipeline as a mechanism for enabling discovery via interactive viewing and measurements of very large collections of images

Image processing pipeline

The many applications of image processing include visualization, information restoration, image retrieval, registration, pattern measurement, object detection, and object recognition.[2] We focus on image processing applications that enable discovery over very large image collections via viewing and quantitative measurements of objects of interest. Scientific discovery is characterized by a workflow (also denoted as a pipeline) of image processing steps. The pipeline is designed by a scientist and depends on the discovery method. Image processing pipelines require human input not only when chaining computational steps but also when selecting computational parameters, choosing information measurements to extract and visualize data, and exploring the semantic meaning derived from image measurements.

Web image processing pipeline

With advancements in microscopy imaging, one experiment can yield large quantities of images that are beyond the processing capabilities of typical personal computers. These processing capabilities include off-line and interactive computations that are critical for making discoveries. We refer to these quantities of images as big image data. In addition, the path to a discovery requires frequent sharing of big image data, intermediate large-scale measurements, and explorations by multiple researchers with varying expertise. These realizations lead us to web image processing pipelines where the processing capabilities (hardware and software) can be scaled to the size of the image data and computational time requirements, while the data sharing and collaborative discovery are supported by the distributed access via web browsers.

---

[2] http://www.engineersgarage.com/articles/image-processing-tutorial-applications

1.2   Web Image Processing Pipeline

In the context of this book, a web image processing pipeline consists of a client-server system and the algorithms that are executed either on the client side or on the server side.

Client-server system
A client-server system can perform computation steps either off-line or interactively (on-demand). For simplicity, the term client can be understood as a web browser running on a researcher's computer or another device. The web browser allows the researcher to view images via computer/device displays and make measurements in the browser environment using the underlying hardware. The term server can be viewed as a web-networked computer (or multiple computers) capable of storing large image collections, serving the images to multiple clients, and handling requests for uploading, computing, and downloading. Data sharing and collaborative discovery is facilitated by the ability of multiple web clients to communicate with a server from any web-networked geographical location. The ability of a server to distribute storage and computational requests from many clients during peak usage to other available computational resources provides the web image processing pipeline with the ability to scale with data size and computational requests. A client-server-based web image processing pipeline requires software algorithms running across diverse web browsers using a server environment.

Client software
Software algorithms running in a web browser are written in JavaScript and are integrated with web technologies such as Hypertext Markup Language (HTML) and Cascading Style Sheets (CSS). While HTML is the language for creating web pages, CSS is the language for describing the style of an HTML document. Unfortunately, existing web browsers do not support the same features of JavaScript, HTML, and CSS languages. Thus, some web image processing steps and interactive features are supported only in a subset of browsers. To address the interoperability, open standards for the web features are critical for the long-term growth of the web and are being addressed by the World Wide Web Consortium (W3C). For example, the W3C HTML Working Group prepared the 5th revision of HTML (HTML5) in 2014 which is used in the web image processing pipeline described in the next chapters.

Server software
The server-side software must communicate with clients and must run user-requested computational algorithms. This communication can be mediated by software such as Apache Tomcat that allows servers and browser-based clients to exchange information similar to a conversation between people. Computational algorithms that are written in multiple programming languages must be seamlessly integrated into the communication between clients and servers. For example, a communication request to stitch a set of overlapping microscopy image tiles into a single image can require the launching of a stitching algorithm with input images and input parameters and the retrieval of the stitching output (i.e., translation vector per

image tile). This assumes that the stitching algorithm has been compiled for the server operating system (OS), can access the input data and output storage, and can be executed in user mode on the underlying hardware [1]. The execution of the same algorithm on distributed and heterogeneous environments requires the addition of software that will distribute the data, schedule the computations, and collect the results.

Client-server communication

The communication and data exchange channel is an important aspect of client-server software design. In practice, this channel is almost always limited by the network bandwidth which has implications for the latency and interactivity of on-demand computations. To overcome latency, especially for very large image collections, images are frequently compressed and use special representations for fast retrieval. For example, large gigapixel images can be represented by a tiled multi-resolution pyramid that is created by iteratively down-sampling the original image size by one half. Each down-sampled image is then tiled into a predefined size (e.g., 256 pixels × 256 pixels). This representation allows for the transmission of smaller more manageable subregions of a gigapixel image (i.e., set of pyramid tiles) for a view or computation.

1.3   Big Data Microscopy Experiments

We will focus on big data microscopy experiments in cell biology and materials science. As mentioned above, these microscopy experiments can yield large image datasets that are beyond the processing capabilities of a typical personal computer. These images are acquired as collections of individual microscope fields of view (FOV). For a fixed acquisition rate, the big image collection originates from imaging a large spatial area at high magnification over extended time and from using many spectral channels (imaging modalities). The image collection also grows with the number of specimens that are sampled from a large specimen bank (e.g., cells from a vial). These "replicate" measurements aim at establishing statistical reproducibility, as well as at understanding response functions under many treatments.

Examples of big data microscopy experiments in cell biology

In cell biology, as well as in histopathology, microscopy imaging provides raw image data about cells and tissues. The research frequently involves studying either population statistics or characteristics of individual cells. The image of an entire specimen is preferable to spot checking randomly selected FOVs for population-level statistics. The spot-checking approach fails to capture rare events or can bias the biological interpretation of cells. There is a need to collect enough data to capture these heterogeneities. Furthermore, in cell therapy applications, it is possible that each cell matters and must be imaged and inspected for quality. Big data microscopy experiments in cell biology are becoming more and more frequent, while the computational solutions for dealing with acquired big data are not yet

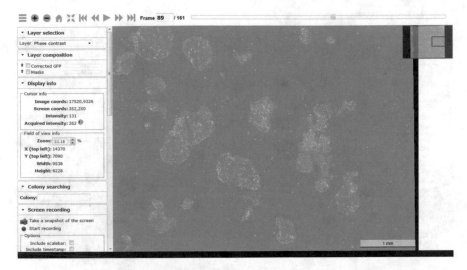

Fig. 1.2 Phase microscopy images of cell colonies imaged over 396 (18 × 22) overlapping FOVs and over a period of 5 days (161 time points at 45 min acquisition rate). The data is available online (https://isg.nist.gov/deepzoomweb/data/stemcellpluripotency)

available. An example of such a big data experiment focusing on cell colony growth and heterogeneity of pluripotency marker is presented in Fig. 1.2.

Examples of big data microscopy experiments in materials sciences

In materials science, microscopy imaging provides raw images about materials and their properties. The research goals include studying properties of materials (ceramics, metals, or polymers), investigating mixtures of materials (compounds, alloys, composites), looking for occurrence of rare particles, or discovering physical and chemical properties of metallic elements and their interactions. The key challenge of these studies lies with the material specimens being much larger than their elements and requiring the imaging of a large area at high spatial resolution. Furthermore, the materials consist of heterogeneous elements that are unevenly distributed and hence spatial sampling must provide sufficient information about the heterogeneity. Figure 1.3 presents an example of a big data experiment focusing on the shape and distribution of aerosolized nanoparticles.

Differences between two application domains

The key difference between the cell biology and materials science application domains lies in the experimental constraints. In order to study living cells, one must choose cell specimen preparation and imaging modalities that do not harm cells and do not change their behavior. Thus, the preferred imaging modality for cell biology is optical microscopy (e.g., bright field, phase contrast, or differential interference contrast). Other microscopy imaging modalities are used to understand cell structure at multiple scales but cannot be used for studying living cells since cells are either fixed

Fig. 1.3 Images of aerosolized carbon nanotubes collected on a Si wafer imaged over 2360 over-lapping FOVs. The data is available online (https://isg.nist.gov/deepzoomweb/data/materialparticlesdistribution)

or destroyed during imaging. Any steerable experiments with living cells also require near real-time processing of images to provide feedback to the experimental configuration (e.g., selected field of view). In order to discover and design new materials in materials science, one is typically more concerned with high spatial and temporal image resolutions and less concerned with destroying the specimen during experiments. In other words, studying material dynamics does not assume that a specimen is unchanged during an imaging experiment while studying living cells assumes that a specimen did not change due to the measurement interrogation during imaging.

1.4   Motivation of Big Data Microscopy Experiments

Measurements at multiple spatial scales

One of the fundamental motivations for conducting big data microscopy experiments is the desire to perform ensemble "bulk" and individual object microscopic measurements. Figure 1.4 illustrates the characteristics of the two types of measurements (top) and the benefits for scientists from big data microscopy experiments (bottom). The measurement problems lie in the large differences between the scales of a specimen and the scale of an observed phenomenon (i.e., cells and subcellular structures are much smaller than the entire Petri dish).

Complexity of studied phenomena

Another motivation for collecting big data is the discovery of complex governing laws, for example, those of biological cells and tissues. Bio- and material

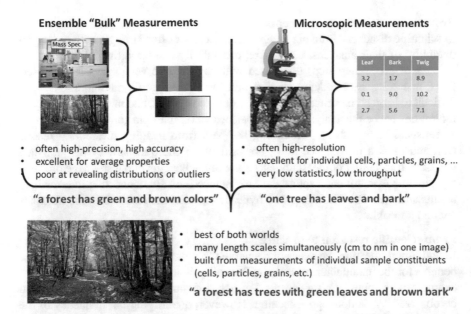

Fig. 1.4 Comparison of today's characteristics of ensemble "bulk" and microscopic measurements (*top*). If solutions for processing big data experiments become available, then a new type of measurement science is enabled

complexity is characterized by many factors, and interactions, and by many temporal and spatial scales. This complexity is related to whole cell and particle responses to various treatments and environments while measuring morphology, function, and spatial and temporal distributions. To study such complex systems, one must collect cell data over the huge space of factors, using many imaging modalities and at many temporal and physical spatial scales leading to big data microscopy experiments.

In addition, the inherent complexity requires leveraging the multidisciplinary expertise of geographically distributed team members and hence the acquired data must be accessible by all team members.

Statistical properties of measurements

By observing stochastic processes changing as a function of many factors, scientists are frequently deriving conclusions from a limited number of observations (i.e., statistical samples of the underlying stochastic process). This raises questions about (a) how many observations to collect in order to derive conclusions with high statistical confidence and (b) which observation to collect in order to acquire representative statistical samples. If statistical samples are chosen based on assumptions about a specimen that do not hold, then validity and reproducibility of data-driven inferences and models (scientific results) are in jeopardy. There is thus a need to run big data experiments to collect as much data as possible to validate appropriate sampling and verify confidence in the conclusions. The statistical benefits of so-called "census" measurements are high and include increased robustness and statistical significance of measurements at the cost of infrastructure investments [2].

Provenance trail of measurements
In scientific discovery from big image data, scientists often struggle with the reproducibility of their analyses. Lack of reproducibility can be attributed to the large number of processing steps associated with algorithmic versions, parameter settings, operating system dependencies, and hardware-specific configurations. It is difficult to keep all intermediate results, software versions, and metadata about configurations hence the analyses may be performed on multiple computers by several scientists. These metadata, describing the path from imaging to reported results (referred to as a provenance trail), are critical for the transparency of published work and traceability of measurements. One of the benefits of a web image processing pipeline is that it can capture the provenance trail. By connecting intermediate results and metadata files using web hyperlinks, scientific results become transparent and traceable.

From scientific research to practice
In addition to meeting the needs of a scientist, big data experiments can have large benefits for the manufacturing sector with quality control and quality assurance. For example, with some luck, rare events during bio-manufacturing of cell therapies can be observed by random spot-checking. However, quantifying the frequency of the rare occurrence requires a well-designed big data experiment, automated processing, and the determination of an overall probability distribution function (PDF). If a rare event has a critical impact, then industries using high-throughput imaging and performing quality assessment may be interested in web image processing pipelines.

Summary of goals of big data microscopy imaging experiments
In a summary, the combination of big data microscopy imaging experiments and web image processing pipelines enable the user:

- To understand multi-scale relationships of phenomena
- To study complex governing laws of phenomena
- To improve statistical properties of measurements
- To facilitate transparency and traceability of research results
- To enable reliable quality control and quality assurance

The ultimate goal for many scientists is also to transition scientific results to bio-manufacturing environments. This transition is much easier if scientific results are accessible and traceable. The web image processing pipeline described in this book is viewed as a tool to make the transition easier. It is also viewed as an open-source component of a cloud-based laboratory information management system.[3] In research labs, it is viewed as an infrastructure that enhances access to expertise and measurement tools.

---

[3] https://appexchange.salesforce.com/listingDetail?listingId=a0N30000008YhTtEAK

1.5   Range of Applications Leveraging Image Processing Pipelines

Applications at the cell level
Live cell microscopy imaging has applications for developing stem cell therapies, for advancing regenerative medicine, and for designing drugs. Whether the cells are administered to the body to benefit the recipient (cell therapy) or the cells are regenerated into tissues and organs to restore normal functions (regenerative medicine),[4] they must be inspected for quality. Microscopy imaging with image processing provides one of the quality assurance tools. Image processing pipelines can be used to monitor cell responses to a large space of treatments during the design process for new therapies, ultimately serving as a replacement for the highly manual and less statistically significant visual spot-checking.

Applications at the tissue level
A web image processing pipeline can help with analyzing histopathology images. These images are collected from biopsies and pose a challenge due to the volume of data, complexity of image content, and a desired short turnaround time on measurements. In comparison to live cell microscopy images, histopathology images are typically stained with hematoxylin and eosin (H&E) and imaged as color images. There is an entire branch of digital pathology devoted to analyzing histopathology images [3]. It is conceivable that digital pathology combined with tailored medical treatment to the individual characteristics of each patient would play a role in an emerging approach to disease treatment and prevention referred to as precision medicine.

Applications at the organ level
Microscopy imaging and image processing have been used to study the brain. Due to the unprecedented size of brain imaging datasets, this application requires sharing data, algorithms, and results, as well as joint collaborations across multiple funding agencies and research institutions. Web image rendering and annotation systems have been the key in the Human Connectome[5] project funded by federal agencies, FlyEM[6] project funded by the Howard Hughes Medical Institute (HHMI) Janelia Farm, and Allen Brain Atlas[7] funded by the Allen Institute for Brain Science.

Materials science applications
A web image processing pipeline can be a tool for material scientists to study images and make measurements of nanoscale particles, for example, that can cause health hazards for exposed humans or failures in engines and building materials. In addition to safety- and quality-related static particle measurements, dynamic measurements are needed to prevent chemical reactions during operations, such as detection of destabilizing chemicals in lithium ion batteries used frequently in cell phones and laptops [4].

---

[4] http://stemcellassays.com/2011/12/distinction-cell-therapy-regenerative-medicine/

[5] http://www.openconnectomeproject.org/ and https://www.humanconnectome.org/

[6] https://www.janelia.org/project-team/flyem

[7] http://brain-map.org/

Other applications generating large size image collections
While this book is focused on microscopy imaging, a web image processing pipeline can be applied to satellite and airborne imaging in geospatial information systems (GIS), as well as to telescopic imaging in astronomy. To adjust to specific measurement objectives of each application domain, image processing steps (algorithms) must be specifically designed, and can be easily integrated into the web pipeline architecture.

1.6   Challenges of Big Data Microscopy Experiments

Given the examples of big data experiments, web image processing pipelines can play a significant role in the transition from big data to knowledge, enabling technologies for real-time control of experiments (i.e., steerable experiments). Extracting knowledge from big image data is challenged by its size and complexity. Steerable experiments are additionally challenged by the limited time for processing, when the imaging microscope requires human input for changing FOVs based on the data it is acquiring. The challenges of big data experiments related to scale, complexity, and speed suggest that web image processing pipelines might offer a viable solution. They hold the promise of managing large image collections and utilizing all investments to acquire big data, sharing the data, providing access for geographically distributed teams, and distributing and accelerating computations that have time-critical aspects. Next, we describe the three main challenges: scale, complexity, and speed.

Data scale
To illustrate changes in data acquisition rates that directly contribute to large image scales, Fig. 1.5 and Table 1.1 summarize the acquisition rates of several microscopes at National Institute of Standards and Technology (NIST). While there are already cameras on the market that can acquire images at a rate of 1 terabyte (TB) per 3 min, our laptops and office/lab desktops are not ready to process at this rate. One can see that one microscope can generate 15 TB in 45 min which is equivalent to all the data in the American Library of Congress that have been gathered since 1800 (i.e., more than 158 million items in 460 languages). The amount of data being generated is currently doubling roughly every 18 months,[8] and the challenge of working at this scale is not going to go away.

Image content complexity
Another property of big data experiments is the complexity of studied phenomena that goes beyond a single discipline, a single expertise, and many times a single institution. The phenomena in cell biology and materials science require a variety of geographically distributed inputs to design image models of multimodal time-dependent TB-sized images over multiple scales. The complexity of image content is illustrated in Fig. 1.6.

---

[8] http://www.datanami.com/2015/08/18/beware-the-dangers-of-dark-data/

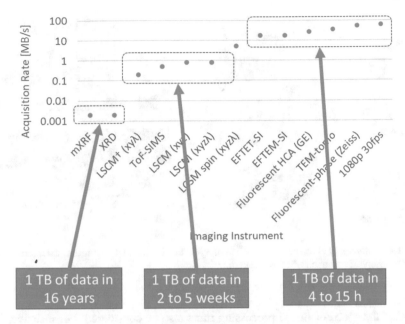

Fig. 1.5 Examples of acquisition data rates by a variety of microscopes

Table 1.1 Microscope abbreviations used in Fig. 1.5

mXRF	Micro X-ray fluorescence spectrometry
XRD	X-ray diffraction imaging
LSCM	Laser scanning confocal microscopy
ToF-SIMS	Time-of-flight secondary ion mass spectrometry
EFTET-Si	Energy-filtered transmission electron tomography-spectral imaging
EFTEM-Si	Energy-filtered transmission electron microscopy
TEM-tomo	Transmission electron microscopy-tomography
1080p 30fps	Wilco Imaging's second generation WIL-HD1080p with 1/3″ Complementary Metal–Oxide–Semiconductor (CMOS) Panasonic sensor, 1080 horizontal lines (progressive scan) and 1920 vertical lines per image (1080p), and 30 frames per second (fps), also denoted as High-Definition (HD) Serial Digital Interface (SDI) camera
Fluorescent HCA (GE)	High content analysis, General Electric (GE) InCell 2200
LSCM spin (xyzλ)	Spinning disk laser scanning confocal microscopy
Fluorescent-phase (Zeiss)	Zeiss Axiovert 200 M fluorescence/live cell imaging microscope

Given TB-sized videos with gigapixel frames, automation of measurements is inevitable. Automated analyses depend on designing models at macro (centimeter) to micro (nanometer) spatial scales over time and imaging modalities. The design of models for algorithms (i.e., segmentation and tracking) relies on limited human visual inspection or other reference measurements. The challenges of this type of complexity require solutions to become interactive so that a computer and a human can work in a tandem.

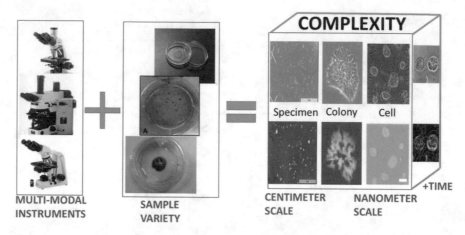

Fig. 1.6 Complexity of image content when imaging cell colonies using multimodal microscopes, over a variety of specimens, in space and time

Speed

Finally, the speed of image processing must also be considered. For instance, if a doctor is depending upon microscope images to monitor a cell therapy being delivered to a patient, and processing that data takes longer than a transition in the state of those cells, there is a risk in delivering the cell therapy. The current computational times on a personal computer are limited to its hardware specifications. For example, after acquiring 2 TB of data in 2 min, it would take (33 to 66) min to move the data over a 1 Gbit/s network, more than 250 laptops with 8 gigabytes (GB) Random Access Memory (RAM) to load 2 TB of data, and about 33 min to perform integer multiplication on an Intel Pentium processor with 3 GHz clock speed.

1.7   Considerations Before and After Digital Images Are Acquired

Specimen condition and image quality

Although this book is focused on the web image processing pipeline after images are acquired, the success of image processing depends very much on specimen preparation and microscope modality and its parameters and calibration. For example, in live cell experiments, a team must make a trade-off between image quality measured by signal-to-noise ratio (SNR) and the amount of illumination that can harm living cells. Higher illumination intensity and acquisition rate is harmful to cells but produces higher SNR images, which makes it easier to detect cells by image processing. Thus, one must compromise to collect meaningful measurements. Furthermore, live cells are fed on a regular basis by exchanging their media. If changing cell media perturbs the imaging configuration, then image registration and cell tracking become difficult.

Microscope configuration

While configuring a microscope, there are also several trade-offs that affect successful image processing. Due to the limited bandwidth for writing to a computer disk, one might have to compromise on the acquisition rate defined by intensity dynamic range (bits per pixel), spatial resolution, and number of channels per time interval. In addition, a traditional microscope is limited by the maximum speed of its motorized stage. Many times, the stability of the specimen and imaging instrument over time is also important for understanding image quality. Finally, the sensitivity of image processing results to microscope calibration has been known and duly noted, for instance, in the case of image stitching [5].

Computational technology

After preparing a specimen, configuring a microscope, and acquiring thousands of FOVs, there are several technology decisions, with related financial costs, to ensure that the necessary computational infrastructure is in place to allow the scientist to explore and obtain measurements for the acquired large data sets. First, raw digital images cannot be visually inspected and analyzed without preprocessing small FOVs (image tiles) into a large FOV. Traditional software libraries used for bioimage processing are not designed for assembling, viewing, and analyzing very large images on desktops and pose a constraint for scientists on extracting useful information from big images. If traditional software libraries for bioimage processing are run in a computer cloud (computer cluster), then they will likely not utilize all cluster nodes and will be limited by the RAM of each cluster node. To overcome this computational constraint, there is a need to facilitate the transition from desktop computing to distributed computing and/or hide this computational aspect from users by designing client-server systems.

Image sharing

Another decision for principal investigators is the allocation of resources between experimental data acquisition, data analyses to learn from acquired data, and data sharing to increase the overall work impact and receive credit for the experimental and data science work. Given significant investments needed for conducting big data experiments, one needs infrastructure for big data analyses and sharing that can help principal investigators to lower costs of needed information technology work. This favorable allocation of resources toward experimental data acquisition can be achieved by reusing open-source solutions and organizing acquired image collections for data sharing. Ideally, big image collections and their analyses can be curated and stored in a data repository and cross-referenced in the corresponding scientific publications.

Collaborative visual exploration and modeling

A principal investigator must decide between the financial cost of a powerful shared server to be used by all collaborators for remote server-based measurements and the lost time and efficiency of having the collaborators download big image data to their local computer for exploration. If the financial resources are available for a powerful server or for provisioning a cloud deployment, then the client-server system can deliver remote "anytime and anywhere" access and customizable, scalable, and flexible computational tools for explorations and modeling.

As an added benefit, server-based image measurements allow for building mechanisms for gathering computational provenance into web systems. The benefit of having computational provenance for traceability of intermediate results and shared images is yet another aspect of web systems that is counterweighted by the cost of software development, hardware to host data, and labor to maintain the web system. Thus, one must weight all aspects of an open source web image processing pipeline including collaborative modeling, data sharing, scalability of computational resources, and computational provenance, against the costs of hardware, additional software, and labor to maintain the web system.

1.8   Enabling Reproducible Science from Big Data Microscopy Experiments

Enabling reproducible science from big data microscopy experiments is a tall order and a lofty goal. We are attempting to make a useful contribution toward this goal by presenting software, data, theory, and practical usage of an open-source web system. This chapter outlined the value of web image processing pipelines in terms of multi-scale understanding of phenomena, studying complex phenomena, improving statistical significance, preparing for high-throughput and time-critical processing, and facilitating data sharing, collaboration, and traceability of research results.

Usefulness of web image processing pipeline
A web image processing pipeline can be viewed as a useful data science infrastructure tool to enable discoveries and facilitate publications from big data experiments. Its benefits to researchers and practitioners lie in increasing scientific productivity and in delivering reproducible results. From a discovery perspective, Table 1.2 summarizes the role of web image processing pipelines in the transition from today's microscopic measurements to tomorrow's macro-to-microscopic measurements. From a publication perspective, advances in the understanding of cell responses under drug treatments can be made more rapidly if an investigator does not have to be concerned about "big data" issues such as configuring computers (the web system is already deployed in the cloud), adjusting software to scale with data size (the software is not limited by the specifications of a local computer), keeping track of intermediate results, guessing algorithmic parameters instead of interactively selecting them, and preparing published materials by linking them with individual data points. A publication with such traceable results also allows reviewers to be more productive and objective because they can now verify results derived from big data collections. This is almost impossible when the data sets are too large and require excessive computation.

Table 1.2 Summary of the role of web image processing pipeline in the transition from today's microscopic measurements to tomorrow's macro- to microscopic measurements from big data microscopy experiments

Role of web image pipeline	Today's microscopic measurements	Tomorrow's macro- to microscopic measurements
Acquire	Small field of view Single length scale	Large field of view Multiple length scales
Measure	Qualitative measurements Image algorithms Questionable traceability Serial processing	Quantitative measurements Well-characterized image algorithms Traceable measurements Parallel processing
Analyze	Spot-checking: Micro-scale measurements Unable to detect rare events Estimated population statistics with large error	Relationship across micro−/macroscales Rare event detection Accurate population statistics
Share and collaborate	Limited data sharing Lack of reproducibility Disconnect between publications and data	Sharing via web Traceability of measurements and analyses Publications linked to data

References

1. Zhou, J.: Getting the most out of your image-processing pipeline. *Techonline*, pp. 1–12, 30-Oct-2007
2. Robinson, C.G., et al.: Automated infrastructure for high-throughput Acquisition of Serial Section TEM image volumes. Microsc. Microanal. **22**(S3), 1150–1151 (2016)
3. Deroulers, C., Ameisen, D., Badoual, M., Gerin, C., Granier, A., Lartaud, M.: Analyzing huge pathology images with open source software. Diagn. Pathol. **8**, 92 (2013)
4. Schalek, R., et al.: Imaging a 1 mm 3 volume of rat cortex using a MultiBeam SEM. Microsc. Microanal. **22**(S3), 582–583 (2016)
5. Chalfoun, J., Majurski, M., Blattner, T., Keyrouz, W., Bajcsy, P., Brady, M.: MIST accurate and scalable microscopy image stitching method with stage Modeling and error minimization. Nat. Sci. Rep. **7**, 1 (2017). DOI: 10.1038/s41598-017-04567-y

Chapter 2
Functionality of Web Image Processing Pipeline

Pipeline functional decomposition

The web image processing pipeline (WIPP) can be viewed as three functional modules that operate on common underlying data. The modules are referred to by their functionality as:

- Web image processing (WIP)
- Web image feature extraction (WFE)
- Web statistical summarization and modeling (WSM)

The modules are illustrated in Fig. 2.1. The WIP module performs field of view (FOV) calibration, image fusion into large FOV images, visualization computations, and object detection. As illustrated in Fig. 2.1, large FOV images along with foreground masks serve as inputs to the WFE module which then integrates widely used libraries for calculating image features and extracts tables of feature values. Finally, the image pyramid (a hierarchical partitioning representation) and the image features are passed to the WSM module which allows (a) spatial- and feature-based filtering of segmented objects and (b) creation of statistical models of histogram summaries linked to the persistent data representation.

Interactions with the three modules

The three functional modules shown in Fig. 2.1 enable imaging scientists to focus on the domain-specific problems (i.e., understanding image content) while using the web-based computational tools to manipulate videos of gigapixel size per frame and frequently reach terabyte-sized volumes. WIPP is web-accessible to facilitate geographically diverse collaborations and to utilize distributed computational resources for CPU-intensive processing. After uploading an image collection, all data are stored in the same database and are accessible by all modules. Users can select the computations and their order of execution while visually inspecting intermediate results, performing parameter optimizations, and following the computational and data provenance traces to reproduce the result.

© Springer International Publishing AG 2018
P. Bajcsy et al., *Web Microanalysis of Big Image Data*,
https://doi.org/10.1007/978-3-319-63360-2_2

In this chapter, we will guide the reader through the deployment of WIPP. We will also describe how to test the deployed software system with the provided test data and how to use the different system capabilities.

2.1   Deploying and Testing the Web Image Processing Pipeline

This section describes the deployment of the three previously described modules (WIP, WFE, and WSM) in WIPP, their associated data, and the workload management system. Two WIPP deployment configurations are briefly discussed here:

- Native installation of the system's modules and dependencies
- Installation using virtualization software on a variety of hardware and software infrastructures

We recommended deploying WIPP using Docker container-based virtualization, which is the focus of this section.

Deployment versus installation
We refer to "deployment" when we push a version of software to several computers (clients and servers) and update the software over time. It is frequently used interchangeably with the word "installation." However, we use the word "installation" in the context of installing a software library onto a single computer and configuring it.

WIPP components
The WIPP modules and their functionalities shown in Fig. 2.1 are enabled by the software components illustrated in Fig. 2.2. The software components have different functional responsibilities described below:

- Web browser/client applications using AngularJS [1] and Web Deep Zoom Toolkit frameworks [2] are responsible for browser–/client-based rendering and

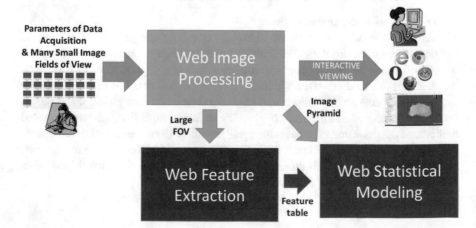

Fig. 2.1 Top-level overview of the three web applications in terms of their functionality

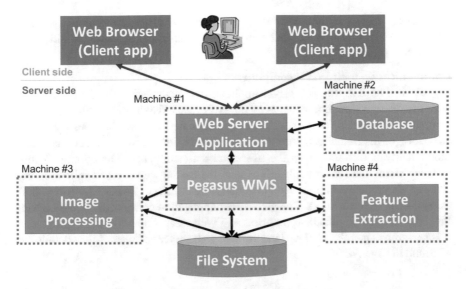

Fig. 2.2 An example deployment of WIPP with color-coded functionality. *Green* – web browsers with user interfaces for the WIPP client application. *Blue* – server computers that perform WIP and WFE computations. *Gold* – data storage consisting of a database for storing all metadata and a shared file system for storing images

interactive measurements. Web statistical modeling is a web browser application.

- Web server application such as Java Spring application [3] is responsible for mediating requests between client and server components of web applications.
- Pegasus Workflow Management System (Pegasus WMS) [4] is responsible for executing computational workflows. The workflows consist of third-party software and NIST libraries with algorithms for image processing and feature extraction. The image processing algorithms are used in the image processing application to perform operations such as stitching, segmentation and tracking of objects of interest. The feature extraction algorithms are used in the feature extraction application to perform feature computations on the regions of interests as detected by the image processing application. Web image processing and web feature extraction modules consist of sets of computational workflows.
- Database such as MongoDB [5] is responsible for storing information.
- File system is responsible for storing images and their image pyramid representations.

These components and technologies are described in Chaps. 4, 5, and 6.

Where to install each WIPP component?

The WIPP software components can be installed on more than one computer. An example of installation on four machines is illustrated in Fig. 2.2. In this example, all web browser communication is handled by computer #1 and is then passed by the Pegasus workflow management system to the WIPP components on machines #3 and #4. Each computer performs computations requested via the web browser and stores its results in both a shared file system and a MongoDB database (installed on machine #2). It is assumed that every user's computer (i.e., the client-side computers) has a web browser.

Communication with WIPP components
As seen in Fig. 2.2, the web server application is responsible for communicating with the other system components, such as Pegasus WMS, the database, and the computation modules. These system components are low-level libraries that facilitate storage, distribute computing, and manage client-server interactions. Their functionality is different from the user-centric functionality. The communication between client and server applications is aided by the web server application and enabled via predefined application programming interfaces (API). The API is built using the Spring framework. The web server application consists of bundled Spring Java application, AngularJS JavaScript, and Web Deep Zoom Toolkit applications. It is served by an embedded Apache Tomcat web server software. More details can be found in Chap. 4.

2.1.1 Types of Deployment

Computer clouds and computer clusters
Web applications are deployed on a set of networked computers. When the networked computers are connected by a local area network (LAN), they are referred to as a computer cluster [6]. Computer clusters are supervised within a single administrative domain and are usually residing in one room. In comparison, cloud computing is a model for enabling ubiquitous, convenient, on-demand network access to a shared pool of configurable computing resources (e.g., networks, servers, storage, applications, and services) that can be rapidly provisioned and released with minimal management effort or service provider interaction according to the NIST definition [7]. In practice, this definition implies that computers might be connected by a wide area network (WAN) and might be highly geographically distributed. Each computer in a computer cluster and cloud is also called a node since the networked configuration can be represented by a graph. The specifications of nodes in a computer cluster are typically the same (homogeneous) while they can vary in a computer cloud (heterogeneous).

Deployment complexity and hardware virtualization
The complexity of deploying a web application on a set of networked computers is greater than that for installing a library on a single computer. There are several methods for packaging web applications and virtualizing hardware that simplify web application deployment. We introduce two of the virtualization methods below. Virtualization allows partitioning of physical resources at a low level so that there is an appearance of having multiple instances of the physical hardware. Computer hardware virtualization can also be viewed as a logical abstraction of hardware components that hides the physical characteristics of a computer from its users.

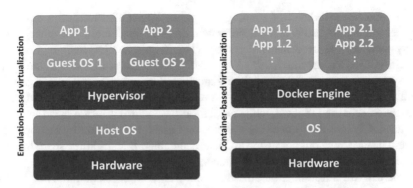

Fig. 2.3 Two types of hardware virtualization. *Left* – emulation-based virtualization. *Right* – container-based virtualization

We consider:

1. Emulation-based virtualization methods
2. Container-based virtualization methods

Figure 2.3 illustrates the differences between these two virtualization methods.

Deployment using virtual machines or emulation-based virtualization method
Emulation-based virtualization emulates a guest operating system (OS) environment running on bare hardware in a host operating system. Microprocessor companies, such as Intel and Advanced Micro Devices (AMD), have added hardware support for virtualization in their commodity processors. This microprocessor support allows entire web applications to be treated as packages with the help of hypervisor software, for instance, VMware Player, VirtualBox, QEMU, Bochs, Parallels, Xen, or Kernel-based Virtual Machine (KVM). In our case, this virtualization method allows us to deploy virtual machines (VM) running a guest OS and WIPP, while the physical machines are running a different OS.

Deployment using Docker container or container-based virtualization
Let us assume that an execution of image processing algorithms requires installations of several libraries with specific versions, manually setting parameters in configuration files, and running several scripts to complete the software configuration before the algorithmic executable is successfully launched. This process must be repeated on each machine and imposes an extra burden on users. If all installations and configurations can be done once and downloaded as a container file, then users can be more productive and the algorithmic executions can be more reproducible [8]. This basic concept of containers is the motivation behind our use of container-based virtualization systems such as Docker [9, 10].

Container-based virtualization systems share low-level resources with the host operating system and are therefore more memory efficient. Applications run in the same operating system as the host. In Linux, this method leverages operating system-level capabilities called Linux Containers (LXS) or runC (formerly known as libcontainer). Linux Containers offer an environment similar to a virtual machine but without the overhead of running a separate kernel and hardware emulation. Docker containers use a layered filesystem (AuFS) by default, but other layered filesystems can be used instead[1] (OverlayFS, Zettabyte filesystem registered by Oracle as ZFS, Virtual filesystem - VFS). AuFS allows common parts of the operating system to be read only and shared among all containers. AuFS also provides each container its own mount point for a writeable filesystem. Container-based virtualization is at the OS level, while emulation-based virtualization is at the processor level. Given a host operating system, one can run a different guest operating system inside a VM but not inside a Docker container. Because they share the same kernel as their host, containers are much smaller than VMs.

One can run different Linux distributions within Docker containers if they use the same kernel as the host. Containers do not require pre-allocating RAM and can be loaded much faster than VMs. These comparative features of Docker container-based virtualization have implications when deploying hundreds of web image processing applications according to varying on-demand requests.

2.1.2   Deployment of Docker Containers

WIPP is distributed for deployments using Docker containers. Links to software downloads, installation instructions, user manual, and test data are provided in the chapter entitled "Supplementary Information Software and Documentation."

Deployment of the WIPP system using Docker containers
We recommend Docker containers for pre-configured and scalable WIPP deployments. The Docker Engine allows for the deployment of multiple container instances to a collection of machines. Docker Swarm allows the containers to form a cluster. It consists of one or multiple manager nodes and worker nodes providing services, as well as an overlay network for multi-host networking. The manager node assigns tasks to the worker nodes in the form of Docker containers that can perform specific services.

To perform a Docker deployment of the WIPP system, use the following steps:

1. Download the ZIP file under the "Docker deployment" section of the WIPP website, containing README files and setup scripts.
2. Install the Docker Engine for running Docker containers on the machine(s) that will host the WIPP system.[2]
3. Create and configure the Docker Swarm for the machine(s) with the Docker Engine. Follow the instructions available both on the WIPP website and in the README files of the downloaded ZIP file.
4. Run the provided setup script to automatically deploy the WIPP system on the created Docker Swarm. Access the system from the Internet Protocol (IP) address of the Docker Swarm manager node.

---

[1] https://docs.docker.com/engine/userguide/storagedriver/selectadriver/#shared-storage-systems-and-the-storage-driver

[2] Follow the instructions at https://www.docker.com/get-docker

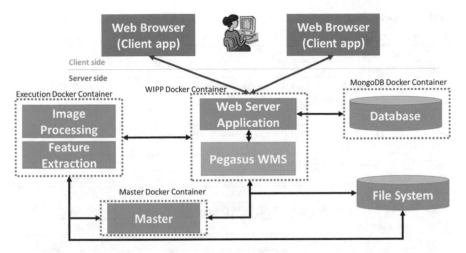

Fig. 2.4 Docker containers that encapsulate components shown in Fig. 2.2

The main components of WIPP are packaged into Docker containers as shown in Fig. 2.4. The "Master Docker Container" is introduced in Fig. 2.4 to coordinate executions using Docker Swarm.

2.1.3   Deployment Recommendations

General recommendations The anticipated image sizes and processing throughput rates should be consistent with the hardware (RAM, disk, and networking). We provide a few general recommendations as well as several computational time benchmarks on a similar configuration (called "test hardware configuration") in Sect. 2.1.4 to guide users during the process of understanding the hardware requirements.

General recommendations:

- Hardware:

  - *CPU and RAM*: Minimum two CPUs and 8 GB of RAM per host for a multi-host deployment or four CPUs and 16 GB of RAM for a single host deployment. Scale up the hardware specifications with the size of input data and the expected speed of processing. We provide a few time benchmarks in Table 2.1. In terms of RAM, segmentation of 1 GB sized image can consume up to 16 GB of RAM. Some other computations are more memory efficient, for instance, flat-field correction, stitching, and pyramid building.
  - *Disk space*: Minimum 50GB of available disk space for small datasets, to be scaled to at least ten times the expected amount of data to be uploaded to the system.

- Software:

  - *Operating system*: Unix family (such as Ubuntu or MacOS)

Table 2.1 Summary of datasets that could be downloaded for testing new deployments

Dataset for testing	Time to process
Background correction	00:00:05
Flat-field correction	00:00:15
Filtering (five types: Mean, median, min, max, Gaussian blur)	00:00:07 per type
Empirical Gradient Thresholding (EGT) segmentation	00:00:04
Fog Bank segmentation	00:00:15
Stitching (four types: Stitching, TIFF stage metadata, no overlap mosaic, time sequence of 1 FOV)	00:00:05 per type
Intensity scaling (two types: Truncation, gamma correction)	00:00:05 per type
Pyramid building	00:00:05
Image assembling	00:00:05
Mask labeling (two types: Four connected, eight connected)	00:00:07 per type
Lineage tracking	00:00:05
Tessellation (square, hexagon)	00:00:05 per type
Feature extraction (two types: Images + masks, images + masks + tiles)	00:00:50 per type

Browser recommendations

For the best experience, we recommend using a recent version of the Google Chrome web browser when using WIPP's web interface. Other web browsers, such as Mozilla Firefox and Apple Safari, are also supported. Some visualization capabilities may not work properly when using Microsoft Internet Explorer or Edge.

2.1.4 Test Data and Computational Benchmarks

WIPP functionality tests and performance benchmarks

Once WIPP has been deployed, users should test the functionalities and collect benchmarks to understand the computational performance of the new system. We provide several test datasets for this purpose. Table 2.1 contains a list of computations and their corresponding time benchmarks. The hardware used for the benchmarks was one virtual machine with four CPUs (Intel(R) Xeon(R) CPU X5670 @ 2.93GHz), 16 GB of RAM, and a mounted network attached storage (NAS) on 1 Gbit/s network. Users can compare the benchmarks obtained on the test hardware configuration against the computational benchmarks obtained with the same data on new hardware or with other datasets relevant to users' scientific domains.

The benchmarks in Table 2.1 correspond to average times. They can vary depending on the network connection and the workload state of the infrastructure while running each computation. The time with "per type" refers to the average time to run a job across the several algorithmic choices ("type").

Provided test data

The test datasets are shipped with the WIPP system and available for download in the "Test datasets" section of the WIPP website.[3] Each dataset comes with input

---

[3] https://isg.nist.gov/deepzoomweb/software/wipp

data, expected output, and a README file containing a set of instructions for configuring and running the computation on WIPP.

Other sources of test data
Another source of test data is the NIST interactive web system.[4] For instance, one can download phase contrast images of cell colonies and use them for testing EGT segmentation, lineage tracking, intensity scaling, and pyramid building. The NIST interactive web system is based on one component of the web image processing pipeline called Web Deep Zoom Toolkit. It is designed for big image dissemination and browser-based measurements (uploading and image processing on the NIST server are not possible).

Additional tests
The use cases described in Chap. 3 provide another way to verify the WIPP functionalities. These use cases are also useful for learning more about WIPP. We will present a few usage examples of WIPP modules in this chapter for testing a WIPP deployment. These examples start with a small stem cell image tile dataset (about 55 MB of data) named "Cy5 dataset" that can be downloaded from the WIPP website[5] under the "Test datasets" section.

2.2 Web Image Processing Module

Top-level functionality
Figure 2.5 illustrates the top-level functionality of the web image processing (WIP) module. A user can upload and process thousands of small fields of view images by submitting computational jobs to the server. The computations are performed on the entire collection of FOV images. Rapid analyses are enabled in a web browser once the small FOV images are processed into a single large FOV image and prepared for zoom-able viewing. The zoom-able viewing and rapid analyses consist of

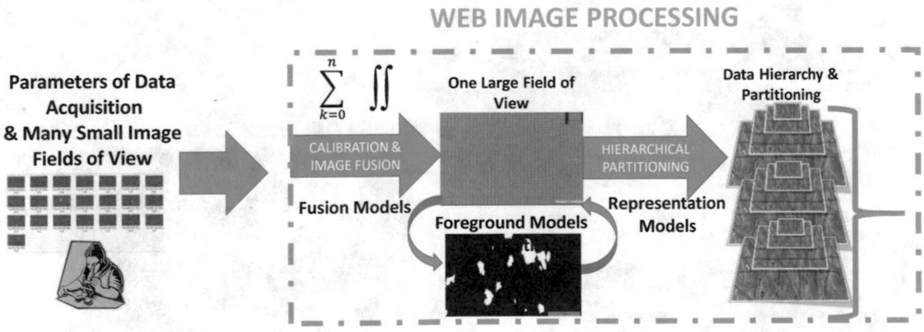

Fig. 2.5 Top-level functionality available in the web image processing module

[4] https://isg.nist.gov/deepzoomweb/data
[5] https://isg.nist.gov/deepzoomweb/software/wipp

user-driven subset selection (like Google Maps) followed by computations over the visible subsets.

2.2.1   *Web Image Processing Module Processing Functionality*

Processing on a server

Web image processing (WIP) module server-based processing creates calibrated, stitched, segmented, and viewable images. It is accessed from a tab called "Image Processing" (see Fig. 2.6) and has the following list of options:

- Image tile calibration (flat-field and background correction)
- Image tile stitching
- Image segmentation (empirical gradient thresholding, EGT, and watershed-based segmentation, FogBank)
- Object tracking across multiple time frames
- Pixel depth conversion to 8 bits per pixel (intensity scaling)
- Image pyramid building for Deep Zoom viewing
- Configuration of multiple image pyramids into a multi-layer Deep Zoom visualization
- Image tessellation to create rectangular or hexagonal image partitions
- Image assembly

Online help

To become familiar with the parameters and user interfaces of each type of server-based computation, we recommend reading the Help section embedded in the WIPP system (accessible from the Help menu tab) or the online user guide on the WIPP website. They provide a visual reference for the dialogs and parameter descriptions.

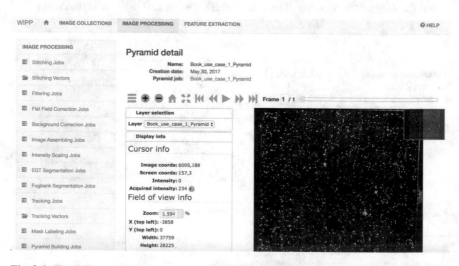

Fig. 2.6 The WIP module user interface for a list of processing operations (*left pane*)

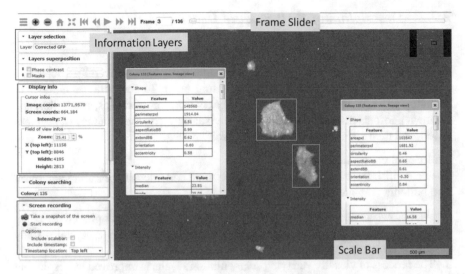

Fig. 2.7 User interface to rapid data subset analyses (*left pane*) applied to viewed data (*main pane*)

Processing in a web browser

The browser-based processing enables:

- Access to calibrated information
- Collection of quantitative measurements
- Testing and optimization of algorithmic parameters
- Support for additional processing by other software tools

The functionality is accessible by selecting "Pyramid" option under the "Image Processing" tab and then viewing the pyramid data. The browser will contain a visualization pane (see Fig. 2.7) with the following list of main options in the left pane:

- Layer selection: choice of pyramid layers
- Layer composition: transparent overlays of multiple layers
- Display information: displaying original and displayed image intensities
- Custom search: colony searching, search for spectral peaks
- Screen recording: capturing in-browser activities
- Data fetching and object fetching: downloading raw or displayed image subsets
- Filtering and connectivity analysis: interactive image filtering and connectivity analysis
- Distance measurements: spatial-scale measurements

The visualization pane functionality is described in detail online.[6] General interactive browser viewing of the underlying gigapixel images uses the icons with tooltips in the left upper corner (see Fig. 2.7) or mouse controls:

- Left mouse click and drag: panning
- Roller spinning: zoom in and out
- Right mouse click: invoke properties of the region if they exist

---

[6] https://isg.nist.gov/deepzoomweb/help

2.2.2 Description of WIP Module Usage

Simple counting exercise

We will now illustrate the utility of WIP module for counting several particles in a 5×5 set of image tiles (denoted as a counting task). To compute the number of particles, process the 5×5 grid of image tiles (flat-field correct tiles, stitch tiles, and segment particles from the stitched image), visually verify the accuracy of segmentation, and then count particles within a specified area.

Additional computations

To accomplish the overall counting task, we must add intensity rescaling and pyramid building computations for four reasons:

1. Web browsers do not support images with more than 8 bits per pixel (BPP).[7]
2. Stitched images are much larger than typical megapixel liquid crystal display (LCD) screens.
3. The process of visually verifying segmentation results is improved by displaying and overlaying both raw image content and detected segments.
4. A region of interest is required to count particles for a given segmentation.

Explanation of workflow steps

Figure 2.8 shows a sequence of computational jobs. The server-side sequence consists of uploading files, stitching and flat-field correction, segmentation and intensity scaling, and pyramid building. Several computations in Fig. 2.8 can be executed in parallel. The sequence yields two overlaid image layers, flat-field corrected raw

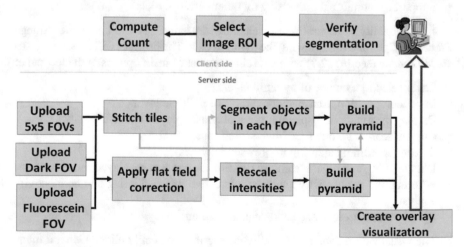

Fig. 2.8 *Bottom* – example workflow to create a zoom-able visualization of flat-field corrected input tiles overlaid with object segmentation. *Top* – three steps performed by a user in a web browser after the overlay visualization has been created. These steps can be accomplished using the web UI

---

[7]Note: microscopes acquire images with 12 or 16 BPP. In addition, computations such as flat-field correction involve multiplication and division yielding 32 BPP images.

images and their segmentation results. These initial computational workflow steps are executed on the server with the remaining steps being performed in the web browser. Segmentation accuracy is visually verified by changing each layer's transparency and verifying the segment alignment with the raw images. Image region of interest (ROI) selection is then performed by panning and zooming by using a mouse or the buttons as shown in Fig. 2.7, top left. Finally, the particle count is computed by selecting the segmentation layer and clicking on the "Run analysis" button in the left pane (Connectivity Analysis segment). This invokes a four-neighborhood connectivity analysis that gives its results as a new dialog with color-coded connected components associated with their area (number of connected pixels). This dialog provides the particle count and size information.

Parameters of the workflow

For this example, here are the required parameters for executing the workflow sequence using the dataset (5 × 5 image tile dataset) described in Sect. 2.1.4, under "module usage examples":

1. Upload the images to a new Image collection: Menu "Image Collections" → "Image Collections" → "Create new collection" → Parameters = [name = Cy5-test-images].
2. Configure and run a stitching job: "Image Processing" → "Stitching Jobs"→"Create new stitching job"→ Parameters are:
   Job name = stitch-Cy5-test-images.
   Tiles collection = Cy5-test-images,
   Algorithm = MIST,
   File name pattern type = Row-Column.
   File name pattern = img_00{r}_00{c}.ome.tif.
   Starting point = Top Left.
   Number of columns = 5.
   Number or rows = 5
3. Segment objects with the EGT segmentation job: "Image Processing"→"EGT segmentation jobs"→create new: parameter = [Job name = EGT-Cy5-test-image, Tiles collection = Cy5-test-images, Min object size = 500, Min hole size = 100, Threshold adjustment delta = 0].
4. Build a pyramid for segmentation: "Image Processing"→"Pyramid jobs"→create new: parameters = [Job name = pyramid-EGT-Cy5-test-images, Stitching vector: stitch-Cy5-test-images, Tiles collection = EGT-Cy5-test-images].
5. Flat-field correct raw images: "Image Processing"→"Flat Field correction jobs"→create new: parameters = [Job name = FF-Cy5-test-image, Raw Tiles collection = Cy5-test-images, Segmented tiles collection = EGT-Cy5-test-image].
6. Scale intensities from 32 BPP to 8 BPP: "Image Processing"→"Intensity scaling jobs"→create new: parameters = [Job name = scaling-FF-Cy5-test-images, Tiles collection = FF-Cy5-test-image, without checking the "Set advanced options"].
7. Build a pyramid for flat-field corrected and scaled images: "Image Processing"→"Pyramid jobs"→ create new pyramid: parameters = [Job name = pyramid-scaled-FF-Cy5-test-image, Stitching vector = stitch-Cy5-test-images, Tiles collection = scaling-FF-Cy5-test-images].

8. Create an overlay visualization of raw and mask images: "Image Processing" →" Visualizations" → create a new visualization: parameters = [Job name = vis-Cy5-test-images].
9. Enter under "Layer group label" =" FF corrected layer", press the plus sign.
10. Enter under "Layer label" = scaled-FF-Cy5, "pyramid" = pyramid-scaled-FF-Cy5-test-image, press the plus sign.
11. Enter under "Layer label" = EGT-segment-Cy5, "pyramid" = pyramid-EGT-Cy5-test-images, press the plus sign.
12. View the overlay of the two channels by clicking on the check box under "layer composition" and moving the slider bar to the right of the EGT-segment-Cy5 label. The web UI for this configuration is shown in Fig. 2.9.

2.3   Web Feature Extraction Module

Top-level functionality

The web image feature extraction module enables extracting traceable image features from raw images and their corresponding segmented masks. The image features include the intensity, spatial, and texture characteristics of objects defined by a 2D segmentation mask. The results are represented as a table of image features and a set of downloadable hyperlinked digital artifacts that were used to compute the image features. Figure 2.10 outlines the top-level functionality (upload → configure and compute → download traceable image features) that uses a range of feature extraction services (i.e., integrated feature extraction libraries). The results contain both the numerical values of image features and the provenance information for the data, algorithms and parameters, software, and computational environment. The traceability of each requested feature is enabled via hyperlinking information as illustrated in Fig. 2.11.

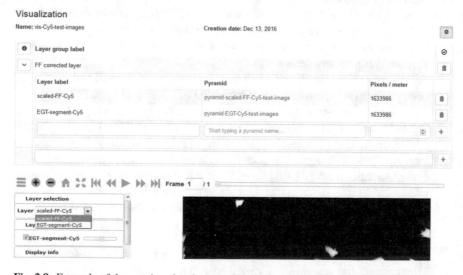

Fig. 2.9 Example of the user interface instance for creating an overlay visualization

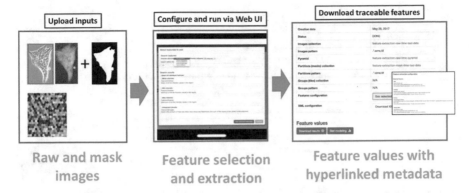

Fig. 2.10 Top-level functionality available in the web feature extraction module

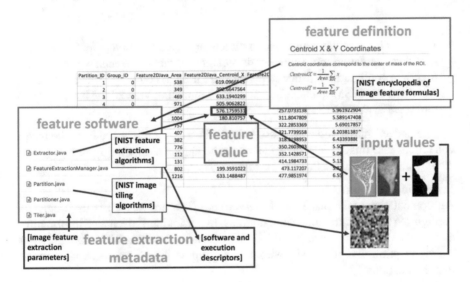

Fig. 2.11 Each feature value (i.e., the value in the middle of the figure in a red box) is traceable to input values, feature definition, feature extraction software and its parameters, as well as to software and execution descriptors

2.3.1 WFE Module Processing Functionality

The WFE module is accessed from the tabs at the top of browser viewing pane (tab "Feature Extraction"; see Fig. 2.12) and includes:

- Selection of image collections
- Selection of image mask collections
- Selection of image mask tessellation collections
- Selection and extraction of image features
- Download of image features as csv files

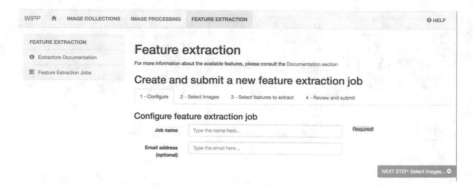

Fig. 2.12 User interface for starting a feature extraction job. The left-side menu tabs provide access to the feature extraction home (introduction page), documentation of available feature extraction packages, and feature extraction jobs page for creation and consultation of jobs

- Access to hyperlinked provenance artifacts of image feature computations (feature documentation and source code, feature extraction executable with input files, and configuration parameters that were used to compute feature values)
- Access to the WSM module for a statistical modeling view of the extracted features (see Sect. 2.4)

The images, mask, and mask tessellation collections are selected from the image collections list. Image collections can either be uploaded into the system or generated by one of the Image Processing Jobs.

Processing location
All computations take place on the server side. The browser UI is used only for configuring feature extraction computations, hyperlinking provenance information, and downloading or viewing results. We will next describe WFE configuration, which is primarily concerned with feature types, image masks, and software implementations.

Configuration of feature types
Feature types describe image objects and can be classified into three types:

- Spectral (intensity)
- Spatial
- Textural

Intensity features are values that are derived from the pixel intensities, such as their central moments (mode, mean, standard deviation, skewness, kurtosis). Spatial features are derived from the locations of pixels, their count, and their locations weighted by intensity value (e.g., perimeter, area, inertia tensor). Textural features capture information about the spatial arrangement of intensities and are computed from second-order statistics (e.g., gray-level co-occurrence matrix) or by transforming images into a space where spatial patterns can be easily quantified (e.g., Fourier transform to quantify spatial periodicity). One can also consider temporal feature

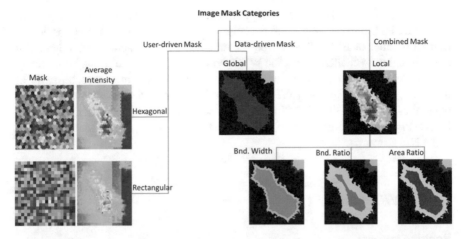

Fig. 2.13 Illustration of image mask categories that define pixel subsets for feature computations

types which describe the evolution of object intensities, textural properties, and spatial coordinates. Temporal feature extractors have not yet been included in the web software system.

Configuration of feature masks
The choice of an image mask is another configurable input. Users can create masks, run segmentation algorithms to obtain a data-driven masks, or combine multiple mask-creation processes. Figure 2.13 shows examples of these three image mask categories. The user-driven masks in Fig. 2.13 are created by partitioning an image into small hexagons or rectangles and computing features over these partitions (also denoted as tessellations). The user-driven masks are useful for exploring spatially varying properties (see the variation of average intensities) or local texture properties that are based on global statistics of features.

A data-driven mask in Fig. 2.13 is a direct segmentation algorithm output. The combined mask examples illustrate how one could analyze the spatial distribution within a data-driven mask by introducing additional boundary width or ratio or area ratio partitioning schemas. These masks are useful for cell colonies containing boundary cells exhibiting different behaviors from those of interior cells because of interactions with the surrounding media. Currently, users can create "combined masks" by uploading data-driven masks (e.g., cell colonies) and a user-driven tessellation mask (e.g., hexagonal partition) and extracting features from hexagons inside of cell colonies.

Configuration of software implementation
The software library and its feature extraction algorithm implementations represent another configurable input. The WFE module integrates many widely used feature extraction libraries originating across scientific fields. These libraries are written in multiple programming languages (Java, Python, MATLAB, C/C++), run on a variety of operating systems, use diverse digital implementations of analog mathematical features, might use different names for identical features, and embed a wide

range of default parameters. The implementations also vary in their use of input image masks (no mask, one mask per region, one mask for all regions, mask with limited number of BPP). This variety of image feature extraction implementation poses integration challenges for the WFE module and raises questions about image feature equivalence across software packages. Image feature traceability is an important advantage of using WFE. At present, we have integrated image features from the following libraries: ImageJ/Fiji, scikit-image, CellProfiler, MATLAB Image Processing Toolbox (regionprops and graycoprops), WndCharm, MaZda, and a NIST Java package.[8]

2.3.2 WFE Module Usage

Ranking exercise

Here we illustrate the utility of the WFE module for ranking cell colonies in a 5×5 set of image tiles (denoted as a ranking task). In this example, the ranking task is to extract the average cell colony intensity as an image feature from the 5×5 grid of image tiles downloaded in Sect. 2.1.4 (step 1) and sorting the intensities from smallest to largest. The task will also include establishing the equivalence of feature names for average intensity across the integrated software libraries.

Input images

We must first prepare the input raw and mask images. The exercise in Sect. 2.2.2 (step 1: segment raw images) gave us an input image collection (Cy5-test-images) and the segmented image collection (EGT-Cy5-test-image). The segmented images contain only two labels, background and foreground (also denoted as the labels of binary images). All cell colonies have the same foreground label in segmented binary images. If the goal of a ranking task is to analyze the cell colonies individually, then each must have a unique label. To assign a unique label to each cell colony, run the mask labeling job in WIP as follows:

- "Image Processing"→
- "Mask labeling job"→
- create new: Parameters = [Job name = label-Cy5-test-images, Images collection = EGT-Cy5-test-image, Connectedness = four connected]

Configuration of WFE

With the labeled mask and raw image collections ready, we switch from the WIP module to the WFE module by clicking the tab "Feature Extraction" menu tab. To configure image feature extraction, we select the tab "Feature Extraction jobs," then "Create new feature extraction job," and type the job name = WFE-Cy5-test-images and an optional email. The specific configuration steps and parameters are shown below:

1. Enter parameters = [Images collection = Cy5-test-images,
   Image names pattern (regular expression)=. *.ome.tif (click on *Pattern help*).

---

[8] https://isg.nist.gov/deepzoomweb/stemcellfeatures#feature-extraction-formulas

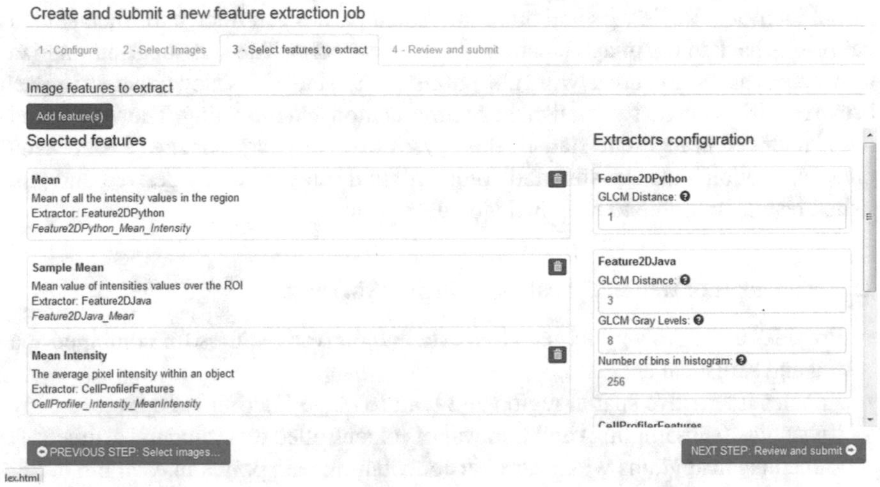

Fig. 2.14 Configuration setup after image features were selected

Check the box "Use Input Partitions (masks)".
Images collection = label-Cy5-test-images.
Image names pattern (regular expression)=. *.ome.tif (click on *Pattern help*).
Follow the tabs at the top or the buttons at the bottom.
2. Choose intensity from "Choose category":
Type "mean" in the search box.
Select the check box next to mean in Python, sample mean in Java, and mean intensity in MATLAB.
See Fig. 2.14.
Follow the tabs at the top or the buttons at the bottom.
3. Review and submit.
4. Once the job is done processing, download the CSV file with the numerical values by clicking on the button "Download results."
5. Load the CSV file into a spreadsheet program and sort all entries based on the column labeled "Feature2DJava Mean."
6. Report the smallest and largest average intensity values (231 and 6448.8) and compare intensity values across columns (i.e., software packages) [identical except from rounding].

2.4   Web Statistical Modeling Module

Top-level objectives
The web statistical modeling (WSM) module enables interactive statistical summarization and modeling of objects of interest found in gigapixel images that are described by image features. These statistical summaries are traceable to each object location in a persistent gigapixel image that is directly viewable using a Deep

Zoom interface. Reviewers and other interested parties can trace published statistical results back to the image location of each data point. The statistical summarization code can be included with the publications since it is downloaded to each browser during interactive statistical summarization and modeling. The WSM module is intended to facilitate statistical analyses over very large images easier (avoid data duplication, data transfer, and computational setup) and any derived and published results traceable to each included data point.

Goals

The main goals of web statistical modeling (WSM) are:

- Provides access to very large image data and derived features for combined spatial and statistical analyses
- Enables interactive spatial region and feature range filtering to support sensitivity studies (subsampling) and removal of uncontrolled experimental artifacts
- Generates histograms with traceable contributing data points in each bin to persistent image collections
- Facilitates visual understanding of feature histograms representing a single object by using image thumbnail of objects
- Provides client–/browser-based functionality to compute statistics, select a suitable model for a histogram to be represented by a probability distribution function (PDF), and estimate parameters of a selected PDF model from the family of Johnson's PDFs
- Saves provenance information about filtering, histogram creation, and statistical modeling in addition to preserving the hyperlinks between histogram data points and their locations in persistent large image collections (i.e., gigapixel images in terabyte-sized video)

Top-level overview of WSM objectives and goals is illustrated in Fig. 2.15 in the context of publishing traceable statistical results.

2.4.1   WSM Module Processing Functionality

Accessing the WSM module

The WSM module serves as a visualization tool for the WFE module's feature extraction job. The configuration of feature extraction job allows for the association of input image pyramids with the extracted features that includes bounding boxes of objects. A histogram of feature values can be presented with the pyramid views of individual objects in each histogram bin. When the feature extraction job is completed, the button "Stat modeling" provides the access to WSM tools as shown in Fig. 2.16.

Top-level functionality

Specific WSM module functions are accessible from the accordion-style user interface on the left side of the browser viewing pane (see Fig. 2.17). These include:

1. Interactive filtering of image objects and their features based on spatial location and feature values (accordion entries "Spatial filter" and Feature filter")

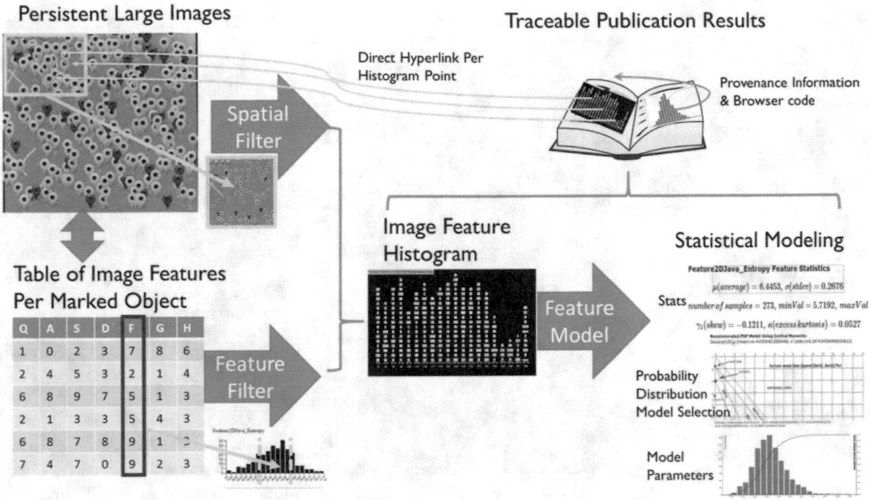
Persistent Large Images
Traceable Publication Results
Direct Hyperlink Per Histogram Point
Provenance Information & Browser code
Spatial Filter
Image Feature Histogram
Statistical Modeling
Table of Image Features Per Marked Object
Feature Filter
Feature Model
Stats
Probability Distribution Model Selection
Model Parameters

Fig. 2.15 Top-level overview of WSM module

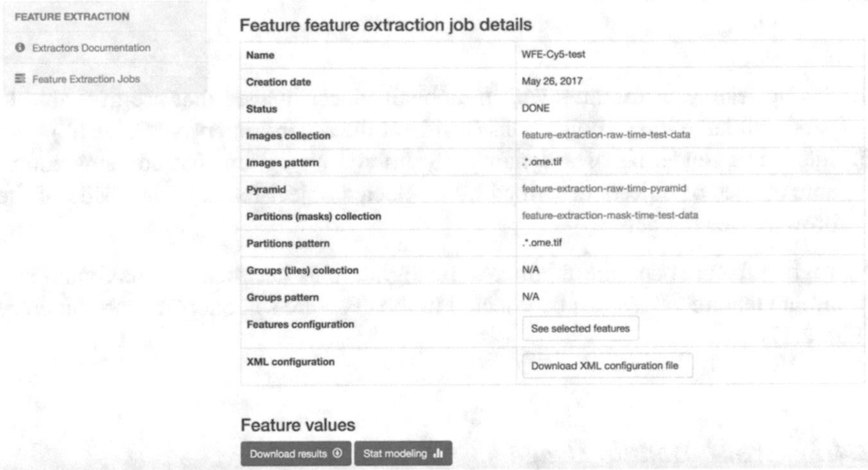
FEATURE EXTRACTION
Extractors Documentation
Feature Extraction Jobs
Feature feature extraction job details
Name
WFE-Cy5-test
Creation date
May 26, 2017
Status
DONE
Images collection
feature-extraction-raw-time-test-data
Images pattern
.*.ome.tif
Pyramid
feature-extraction-raw-time-pyramid
Partitions (masks) collection
feature-extraction-mask-time-test-data
Partitions pattern
.*.ome.tif
Groups (tiles) collection
N/A
Groups pattern
N/A
Features configuration
See selected features
XML configuration
Download XML configuration file
Feature values
Download results
Stat modeling

Fig. 2.16 Results of image feature extraction job include a computational setup (general information) and numerical values (*bottom*: download of feature values, access to statistical modeling view)

2. Computing statistics of filtered features (inside of the accordion entry "Statistical modeling"→"Show Stats")
3. Estimating statistical probability distribution function (PDF) models from image feature histograms and generating feature values according to the estimated PDF model (inside of the accordion entry "Statistical modeling"→"Recommend PDF model" and "Estimate PDF parameters")

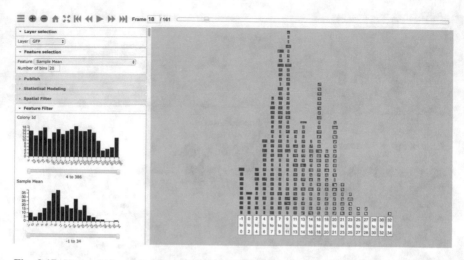

Fig. 2.17 User interface to web statistical modeling. *Left* – accordion entries for selection of parameters such as a histogram feature and the number of histogram bins that divide the maximum and minimum feature values, spatial and feature filtering, statistical modeling, and publishing. *Right* – dynamically filtered histogram that contains zoom-able images of objects described by a selected object feature

4. Saving feature histograms with thumbnail object images that are traceable to persistent large image collections (inside of the accordion entry "Publish")
5. Interactive rendering of a dynamically filtered histogram that contains zoom-able images of objects described by a selected object feature (right side of the browser viewing pane)

Histogram features and the number of histogram bins between the maximum and minimum feature values can be selected in the "Feature selection" accordion entry (Fig. 2.17).

2.4.2   WSM Module Usage

Traceable histogram exercise

We will illustrate the utility of WSM module for estimating a histogram distribution of average cell colony intensity in a 5 × 5 set of image tiles (denoted as a histogram task). The histogram task assumes that an average cell colony intensity was extracted using WFE (Feature2DJava_Mean feature) from raw and mask images created in WIP module as previously described. The histogram task includes filtering values smaller than 500 and larger than 6000, estimating statistics of filtered histogram data points, and saving the traceable histogram.

Inputs

The inputs for the WSM module are image object features and the corresponding image pyramid. By default, the statistical modeling view will be automatically generated based on the selected pyramid and extracted features while creating a feature extraction job during the input image collection configuration. The histogram view on the right panel of the WSM module is dynamically drawn when changing the features or number of bins or when applying spatial and feature filters.

Interactive steps in WSM

With the inputs set, we can interactively generate results for the histogram task using the following steps:

1. Click on "Feature selection" in the accordion UI on the left side and select the Feature2DJava_Mean feature to generate the corresponding histogram on the right side.
2. Click on "Feature filter" in the accordion UI on the left side and scroll down to the Feature2DJava_Mean feature histogram.
3. Move the slider bar under the histogram on the left side to the right to eliminate data points less than 500.
4. Move the slider bar under the histogram on the right side to the left to eliminate data points larger than 6000.
5. Click on "Statistical modeling" in the accordion UI on the left side and then on the "Show stats" button. Copy the values in the pop dialog.
6. Click on "Publish" in the accordion UI on the left side and then on the "Histogram snapshot" button.
7. Download the zip file, extract the histogram files, and verify the HTML hyperlinks associated with each thumbnail image connects to the objects in the image pyramid.

Public access to the demonstration of WSM

The WSM module is also deployed for data exploration on the publicly accessible website[9] via the "Statistical modeling" link under each of the replicate dataset. In this open access WSM module instance, co-registered images were selected from the stem cell experiment conducted at the Material Measurements Laboratory at NIST using phase contrast (PC) and green fluorescent protein (GFP) microscopy imaging (half a gigapixel per image, 16 BPP). Each image captures hundreds of automatically segmented stem cell colonies. Each colony is described by 78 features that are potential indicators of stem cell health. The goal is to discover population-level statistical models and rare events and to assess sensitivity of statistical models to spatial location. The demonstration allows scientists to perform browser-based statistical analyses of all features from two imaging channels and to publish the results of statistical analyses that are traceable to the persistent Deep Zoom pyramid visualization.

---

[9] https://isg.nist.gov/deepzoomweb/data/stemcellpluripotency

2.5 Summary

This chapter introduced three functional modules of WIPP. The modules allow users:

1. To explore thousands of FOV images acquired during a big data microscopy experiment
2. To characterize objects of interests with a variety of measurements
3. To model the characteristics with probability distribution functions over large object populations in time

Once images are converted into object measurements, the data size is typically significantly reduced. Users can then download the files and proceed with other tools that provide more functionalities but have not been designed for large dataset.

References

1. "AngularJS," [Online]. Available: https://angularjs.org/. [Accessed: 25 Sep 2017] (2017)
2. Bajcsy, P., et al.: Enabling interactive measurements from large coverage microscopy. IEEE Comput. **49**(7), 70–79 (2016)
3. Walls, C.: Spring in Action: Covers Spring 4, 4th edn. Manning Publications, Saintmpford (2014)
4. Talia, D.: Workflow systems for science: Concepts and tools. ISRN Softw. Eng. **2013**, 15 (2013)
5. Chodorow, K.: MongoDB: The Definitive Guide, 2nd edn. O'Reilly Media, Sebastopol (2013)
6. Baker, M.: Cluster computing white paper, University of Portsmouth, UK (2000)
7. Mell, P., Grance, T.: The NIST definition of cloud computing recommendations of the National Institute of Standards and Technology. NIST Spec. Publ. **145**, 7 (2011)
8. Silver, A.: Software simplified. Nature. **546**(7656), 173–174 (2017)
9. Turnbull, J.: *The Docker Book: Containerization is the New Virtualization*, Kindle. Amazon Digital Services LLC (2014). https://www.amazon.com/Docker-Book-Containerization-new-virtualization-ebook/dp/B00LRROTI4#reader_B00LRROTI4
10. "Docker," [Online]. Available: https://www.docker.com/what-docker. [Accessed: 25 Sep 2017] (2017)

Chapter 3
Example Use Cases

3.1 Cell Count and Single Cell Detection

The task in this use case is to quantify the cell count in a well. To obtain the cell count measurement, we will discuss segmentation of single cells from green fluorescent protein (GFP) images and then utilize the histogram of cell sizes to understand the cell count inaccuracy due to the objects containing more than one cell in them.

Challenges

Cell seeding on a plate is a common practice in many laboratories. The operator has limited control of the cell placement leading to a frequently random spatial distribution of cells. Within (24 to 48) h after cell seeding, cells are often in contact. Segmentation techniques can detect single cells in images with higher accuracy and confidence if cells are well separated. The confidence in the cell count measurement decreases when the cells in a FOV are touching.

Inputs

We will analyze one well on a plate that is randomly seeded with A10 cells. The well is imaged using a phase contrast microscope as a grid of (23 × 29) tiles with 10 % overlap. Each tile has a dimension of (1392 × 1040) pixels with intensities represented by 16 bits per pixels (BPP). Each pixel dimension is equivalent to 0.644 μm (i.e., 10× magnification).

Analyses

After uploading images to WIPP, we begin by stitching the single image tiles into a large FOV image and then use segmentation to detect all cells in the well. Next, we identify the single cells from a group of cells. Once we identify cell locations in the well, experts can perform additional manual validation of the results.

© Springer International Publishing AG 2018
P. Bajcsy et al., *Web Microanalysis of Big Image Data*,
https://doi.org/10.1007/978-3-319-63360-2_3

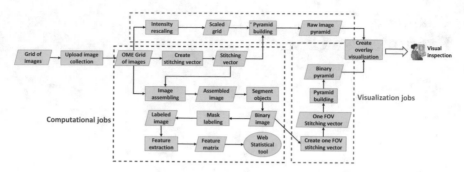

Fig. 3.1 Image processing workflow for computing a cell count measurement and for identifying areas in the image with high and low cell count confidence

3.1.1 Image Processing Workflow

The image processing workflow to extract cell count consists of the following steps:

1. Create a new collection and uploading images.
2. Stitch image tiles.
3. Scale intensities and build an image pyramid.
4. Assemble tiles into a large FOV image.
5. Segment.
6. Extract image features.

Figure 3.1 displays a detailed image processing workflow to solve that the problem of cell counting. The items marked in long dotted orange lines are for visualization purposes only. Please note that the intensity scaling is applied only for visualization, while the image assembly is applied to the raw input tiles. Next, we describe how to input correct parameters for each workflow section.

3.1.2 Create a New Image Collection

After acquiring microscopy images, all files are stored on a disk. The user can upload the files to WIPP by following the next steps:

1. From the "Main Page," click on "Image Collection" tab.
2. Access the "Manage Image Collection" page.
3. Press the "Create new collection" button and enter the name of the dataset. This name will be saved and tagged to that dataset.
4. Press the "add files to collection" button and browse for the saved files or you can drag and drop the files into the browser area.

> **Note**
> Add as much metadata as possible in the dataset name. The name of the experiment and the acquisition date are the most common metadata naming scheme.
> Descriptive dataset names will make searching for them easier at later times.

This is a large dataset and therefore the upload will take some time to finish. Once the upload is completed, the system will convert the files to the open-source ".ome" format (Open Microscopy Environment). We recommend checking the files in the browser under "Collection details" after the conversion. After a successful upload, the lock button must be pressed before any processing can be applied to the image collection. Locking the collection prevents deletion from the system to preserve computational provenance.

> **Note**
> Once the dataset is locked, it is available to the algorithms (or jobs) as input. Jobs are accessed from the "Image Processing" tab.

3.1.3   Stitching of Image Tiles

Since we are interested in the entire imaged well, we must find the exact position of a set of individual FOVs in the coordinate system of the large FOV by performing stitching. The stitching algorithm relies on the microscope acquisition parameters that are either embedded in the image files or provided by the user. For this example, the acquisition contains 16×22 FOVs (image tiles) acquired at 10% spatial overlap along both horizontal and vertical directions. The acquisition starts with the upper left tile and continues by moving horizontally in a combing pattern. In this experiment, the microscope software named the image tiles sequentially with regard to their position, time, and channel.

> **Note**
> For acquisition, some users choose a sequential file naming style, while others may choose a row- and column-based naming style. The stitching job in WIPP can handle both naming conventions.

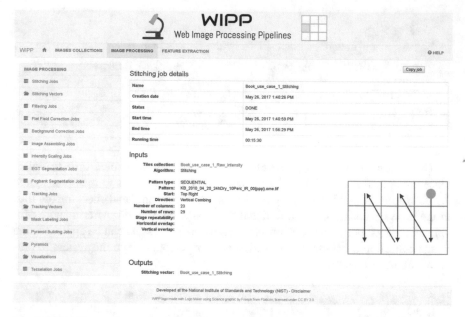

Fig. 3.2 Stitching output

The steps needed to run the stitching job:

1. Click on "Image Processing" and select "Stitching job."
2. Click on "Create new job."
3. Enter the parameters into the web form invoked by "Create new stitching job" (from Fig. 3.2).
4. Run the stitching job.
5. Wait until the algorithm displays "Done" when finished.

The output of the stitching job is in the "Stitching Vectors" page accessed from the "Image Processing" panel.

3.1.4 Intensity Scaling and Pyramid Building

Most microcopy images are in the 16 BPP (uint16) format. Web browsers are able to only render eight BPP images and require scaling the original images before launching the pyramid building to enable visualization of the large FOV image. This is accomplished in the following steps:

1. From the "Image Processing" panel, select "Intensity Scaling Jobs" and create a new job.
2. Select the Raw collection, the default parameters, and then launch the job.

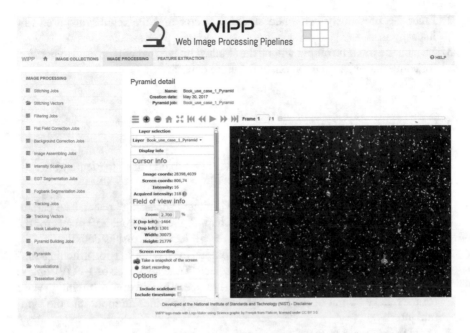

Fig. 3.3 Pyramid image of the A10 cell plate experiment

3. When the job is done, go to "Pyramid building job" and click on "Create new pyramid job."
4. Input the job name, select the stitching vector and the A10 cells grid collection, and then launch the Pyramid building job.
5. After the pyramid is built, the user can view the large image by clicking on "Pyramids" in the "Image Processing" panel and select the newly created job (Fig. 3.3). This makes it easier to inspect the image and visually check the stitching performance.

> **Tip**
> Use the left mouse button to pan around the image and the scroll wheel to zoom in and out.

3.1.5   Image Assembling

Quantitative analyses are performed on the original raw intensity images. We need to assemble the large FOV image before segmenting the cells. The FOV assembly is performed using the following steps:

1. Select "Image assembling job" in the "Image Processing" panel and select "Create new job."

2. Enter the previously computed stitching vector and the regular acquired raw intensity grid.
3. Launch the job. The output will be the stitched image that will be used to perform for computation, segmentation, and feature extraction.

3.1.6 ⸰ Segmentation

Now we can proceed to segmenting the newly assembled image to identify the cells or groups of cells in it as follows:

1. Select "EGT Segmentation Jobs" in the "Image Processing" panel. The parameters of EGT are described in detail in Chap. 5. We will input 250 pixels for the "minimum object size" parameter, and we will leave the rest as default.
2. Visualize and verify the segmentation output by navigating to "stitching job" from the "image processing" panel and select "Time sequence of 1 FOV" option from the algorithms drop-down menu. This operation creates a stitching vector for the binary output of EGT segmentation and becomes an input into the pyramid job.
3. Build the pyramid of the binary image.
4. Visually inspect the segmentation under "Visualization" in the "Image processing" panel.
5. Select "Create new visualization" and enter a name. Under "layer group labels," input "Layer 1" and select the "+" sign to add layer 1 to the visualization job. Input "Layer label," i.e., "raw intensity pyramid," select the pyramid of the raw intensity by typing the corresponding name, and click on "+." This will create the first layer to visualize. Repeat the process by adding a second layer for the binary image.

Once the visualization is created, use the slider bar to change the transparency of the two layers and visually check the segmentation result (Fig. 3.4). Visual verification of segmentation results is always a useful step in analyzing large images.

3.1.7 Binary Image Labeling

The output from EGT is a binary image with the label set to 1 for all foreground pixels and another label set to 0 for all background pixels. To find the location of each segmented object (cell), we need to run the "Mask labeling job" under the "Image processing" panel. This operation assigns a unique label to image regions that contain pixels connected either by four or eight neighbors (four or eight connectivity).

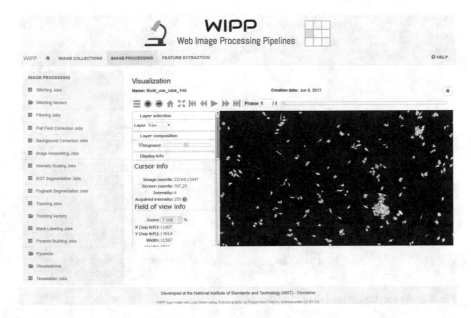

Fig. 3.4 Visual inspection of the segmentation results

3.1.8   Feature Extraction and Single Cell Detection

Because we are interested in characteristics of each cell segment, we run the feature extraction job by clicking the "Feature extraction" tab and perform the following steps:

1. In the "Feature extraction" panel, click on "Feature extraction job."
2. Create a new job and assign a name to it. This will take us to the image feature extraction job creation.
3. Under "select images," select the dataset we are working on, and then check the box that states "pyramid-optional." This option will allow the feature extraction job to generate inputs for using the web statistics tool.
4. In the feature extraction job, click on "add feature" under "select features to extract." From the list of features tool, select the "Java" option under "extractor" and the "Area" as the feature to compute.
5. Click on "Stat modeling" button and select "Area" as the feature to analyze.

3.1.9   Discussion

Figure 3.5 presents sorted cell areas in the large FOV image. One can visually choose the area threshold beyond which a cell is considered a group of cells rather than a single cell. The confidence in detecting isolated cells is higher than those in contact with others. By finding spatial regions with groups of cells, the user can simply ignore them from analysis or analyze them visually.

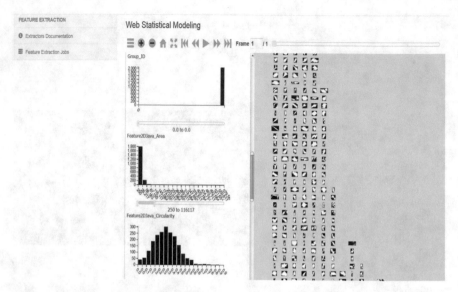

Fig. 3.5 Web statistical modeling of A10 cells

Filtering the results of high-confidence regions of interest is very useful for big data microscopy experiments with multiple variables and complex experimental setups. It increases the confidence in the results being analyzed to produce better measurements. The measurements can include cell counts, average cell intensity, or spatial-textural information about protein expressions in cells.

3.2   Stem Cell Colony Growth Computation

The task in this use case is to quantify the growth of stem cell colonies through time.

Challenges
Pluripotent stem cells exist in a privileged developmental state with the potential to become any cell type in the adult body. Pluripotent cells grow as isolated colonies with each colony comprised of tens to thousands of cells as the culture progresses in time. The colony growth is an indicator of the cell population health. Since the spatial extent of individual colonies is much larger than the spatial coverage of a single camera field of view (FOV), colony tracking can only be done from movies of large FOV images.

Inputs
In this case, we made a movie from a 16×22 grid of individual FOVs (total size of one large FOV image ≈ 1 GB) with a 10% spatial overlap between adjacent FOVs in both X in Y directions. Images were collected over time using phase contrast imaging. The actual experiment generated data from 5 days at a rate of one 16×22 grid of FOVs every 45 min for a total of 161 mosaicked grids. In this example, we apply WIPP to the first ten mosaicked grids (large FOV frames).

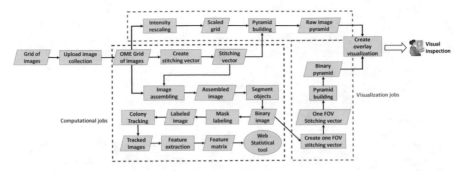

Fig. 3.6 Pipeline for stem cell colony growth computation

3.2.1 Image Processing Workflow

The image processing workflow to extract colony growth consists of the following steps (Fig. 3.6):

1. Create a new collection.
2. Stitch image tiles.
3. Scale intensities and build an image pyramid.
4. Assemble the image.
5. Segment.
6. Track segments.
7. Extract image feature.

The first five steps of the image processing workflow are identical to the ones in the first use case documented in Sect. 3.1. We will assume that the reader has already uploaded the images, stitched them, and then segmented them. We focus on steps 6 and 7 to complete the dynamic measurement (tracking).

Figure 3.7 shows the stitching parameters for the stem cell data experiment. The dataset is acquired as a sequential time sequence with a grid containing 16 columns and 22 rows and the acquisition starting in the upper left corner of the grid and along the horizontal combing direction.

3.2.2 Colony Tracking and Feature Extraction

The tracking job associates colony identifications (IDs) with the segments as they move in time. We will start by performing tracking on the segmented collection:

1. Launch the tracking job by clicking on "Tracking job" in the "Image processing" panel.
2. Enter the parameters as shown in Fig. 3.8 and launch the job. The tracking takes labeled images in input and outputs a set of globally labeled images where each unique region of interest will have a unique label assigned to it across the time sequence.

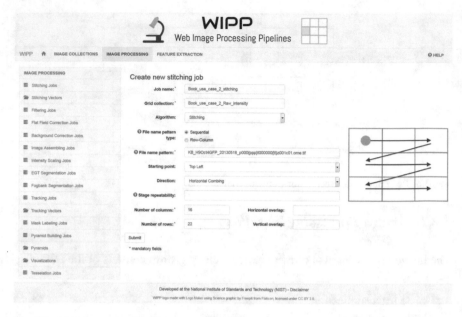

Fig. 3.7 Stitching parameters for stem cell dataset

Fig. 3.8 Tracking parameters

3. Download the output of tracking and explore the images and the associated output matrices.

It is important to note that colonies grow by proliferation (mitosis) and by merging with other colonies over time. The colony mergers lead to having multiple colonies at time t and a single colony at time $t+1$ because of a merger. When a merger occurs, the WIPP tracker (Lineage Mapper) identifies the event as a colony fusion and reports it in the fusion matrix (one of the tracking outputs). The WIPP tracking algorithm can create a fusion lineage tree as well as the regular mitosis tree. After colony fusion, the fused colonies lose their identities, and the newly formed colony is assigned a new label. Once the tracking is completed, the colony area is computed as in the first use case. The only difference is that the unique identities associated with segments (cell colonies) are tied via tracking.

3.2.3   Discussion

The output of feature extraction is ten files in the CSV file format with tabulated information for each colony per time frame. The reader can post-process these files to extract the colony growth over ten time frames. We grouped these ten individual files into one file to obtain one row per colony with area values sorted by time along columns and used that file to plot the results. The post-processing code in MATLAB is provided below:

Post-processing code: Transformation of 10 tables with area values

```
% Define CSV file location on disk
xls_file_loc = '\\Your\csv-files\';

% Read all CSV files
files = dir([xls_file_loc '*.csv']);
% Compute the number of files (=10 in this example)
nb_files = length(files);
% Initialize the output area matrix
A = zeros(300,nb_files);
% loop through the files, read the content and save the
area of each colony in the corresponding matrix
for i = 1:nb_files
    % Read CSV file content
    F = xlsread([xls_file_loc files(i).name]);
    % Save information in matrix A
    A(F(:,1),i) = F(:,3);
end

% Set all zero values to nan (it won't show in the plot)
B = A;
B(A==0) = nan;
% Plot the log10 of area
figure, plot(log10(B)')
```

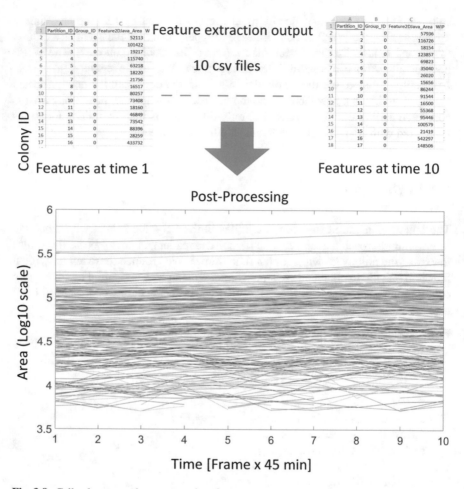

Fig. 3.9 Cell colony growth across ten time frames

After post-processing, one can visualize the area as a function of time for each colony (the last command in the post-processing code). Figure 3.9 displays the area of all colonies through ten time points separated by 45 min. Cell colonies are expected to double in size every 24 h.

The exponential growth can be determined as:

$$Area(n+1) = Area(n)^* 2 \tag{3.1}$$

$$Area(t) = Area(1)^* 2^{(t^* 45)/(24*60)} \tag{3.2}$$

where n is the number of days and t is the time frame index. The exponent considers acquisitions 45 min apart converted to days. Given the vertical log10 scale in Fig. 3.9, there is a lot of noisy data for small colonies. There is a total of 198 colonies in time frame 1. Following the formula above, we computed the theoretical area at time $t = 10$ for all 198 colonies. Next, we computed the relative difference between that theoretical area and the actual area measured at $t = 10$ for all colonies larger than 50 000 pixels in size. There is a total of 90 colonies with the area size above that threshold. The mean value of the relative difference across all 90 colonies is -0.0875 (8.75 %), and the standard deviation is 0.1589. These results suggest that colonies are growing at the expected growth rate.

Figure 3.10 shows the relative area difference between measured and theoretical area values for the ten frames and for all colonies. The colonies whose size is less than 50,000 are not being displayed. We can see in Fig. 3.10 that the relative difference for most colonies is between ±0.2. Some portion of that difference is due to segmentation errors around colony borders (pixels might not have been included). However, six colonies display smaller growth rate than the rest. We identified these colonies' IDs to be 27, 99, 119, 158, 165, and 192, and one of them is being displayed in Fig. 3.11. We can keep a closer look at these colonies and analyze their growth beyond the current small sample of ten frames. Will their growth rate keep decreasing through time (maybe a sign of dying stem cells) or will it bounce back and go back to normal throughout the 24 h? The reader is welcome to explore the entire collection of 168 frames that can be found at isg.nist.gov and find the answers to these questions.

The following is the MATLAB code to compute these statistics:

```
% Compute the area growth for large colonies
% Get the size of all colonies at t=1 and t=10
A1 = A(:,1);
A10 = A(:,10);
% Set the no colonies index to nan
A1(A1 == 0) = nan;
A10(A10 == 0) = nan;
% Compute theoretical area given a double size rate every
24hrs
A10t = A1*2^(45*10/(60*24));
% Compute relative difference between measured and
theoretical area
d = (A10-A10t)./A10;
% Filter results based on 100k area size
ind = A1>50000;
d(~ind) = nan;
% Plot relative difference and compute mean and standard
deviation
figure, plot(d,'.')
mean(d,'omitnan')
std(d,'omitnan')
```

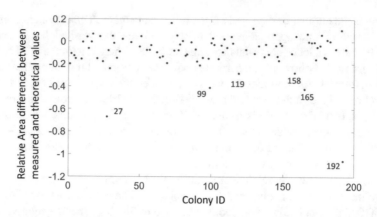

Fig. 3.10 Relative area difference between measured and theoretical area values after 450 min

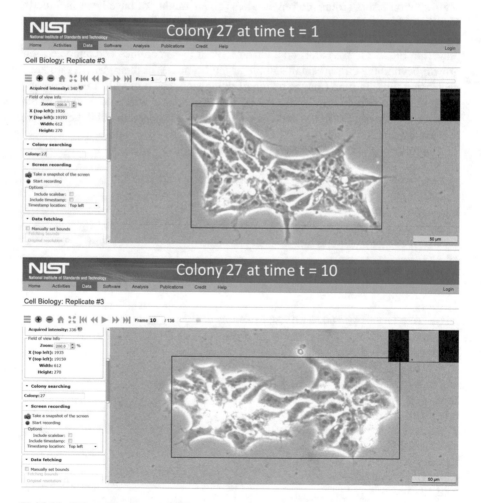

Fig. 3.11 Colony 27 at times 1 and 10

3.3   Image Feature Variability and Its Impact

The task in this use case is to quantify the image feature variability. In this section, we will analyze numerical variability of image features extracted using multiple feature extraction software packages. The use case presented here was selected to shed some light on the importance of measurement science.

Challenges

Can we obtain the same numerical values of image features in multiple labs? There are many ways to approach this question since the number of potential error sources is very large across laboratories. One of the approaches is to investigate the use of different software packages for computing the same image features. We will assume that all other variables in multiple labs would not contribute to the image feature variability.

Inputs

We will be working with the live phase contrast 3 T3 images comprised of 238 images and a total of 8162 cells with different shapes and sizes (Fig. 3.12). We will assume that the reader already knows at that stage how to upload, segment, track, and compute features on these images.

3.3.1   Image Processing Workflow

The image processing workflow to extract image features consists of the following steps:

1. Create a new collection.
2. Stitch image tiles.
3. Scale intensities and build an image pyramid.
4. Assemble the image.
5. Segment.
6. Track segments.
7. Extract image feature.

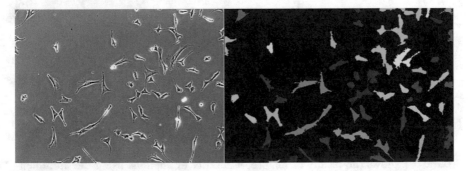

Fig. 3.12 Example test image (*left*) and its corresponding segmented mask (*right*). Each ROI in the segmented mask has a unique randomly chosen color for display purposes

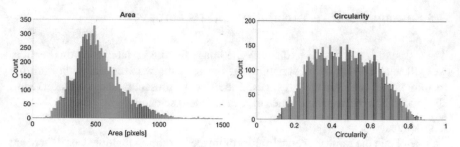

Fig. 3.13 Histograms of area (*left*) and circularity (*right*) features from the objects defined by test images and their masks

In step 7, select all four software packages with the total of 218 unique features. The subsets of unique features are implemented in Python (40 features), ImageJ/Fiji (33 features), Java (74 features), and CellProfiler (101 features). Python features were implemented on top of an existing image processing library (scikit-image [1]), ImageJ/Fiji [2] features were implemented as a plugin using the ImageJ application programming interface (API), and Java features were implemented from scratch at NIST [3].

3.3.2   Image Feature Variability Analysis

We will focus primarily on variability of intensity and shape features. Figure 3.13 shows the histogram of test image measurements for area and circularity features.

Evaluation Metric
Given two vectors of feature values V_1 and V_2 computed over a set of ROIs (image segments) by two software implementations of the same feature, we compute their dissimilarity metric S as the sum of relative errors E_i^m normalized with respect to the average of the two values from the vectors V_1 and V_2 that exceed a given threshold:

$$S = \sum_i \left(E_i^m > T \right) \tag{3.3}$$

$$E_i^m = \left| V_{1i} - V_{2i} \right| / \mathbf{mean} \left(V_{1i}, V_{2i} \right)$$

where $i = 1,\dots,n$ and n is the number of ROIs. T is the user defined error threshold defined as 1 % of E_i^m in this work. The purpose of T is to detect substantial feature differences. The error is normalized by the average value.

Image Feature Variability Analysis
Table 3.1 shows the results of feature variability evaluations using the aforementioned metric. The "Agree" column indicates when software have less than 1 % error across all 64 test cells. The "Disagree" column indicates whether there is an error

Table 3.1 Summary of common feature variability between tools based on metric S (I = ImageJ, J = Java, P = Python)

Feature name	Agree	Disagree	Absent
1. Perimeter		P, I, J	
2. Solidity		P, I	J
3. Circularity		I, J	P
4. Skewness		I, J	P
5. Kurtosis		I, J	P
6. Centroid_X	P, J	I	
7. Centroid_Y	P, J	I	

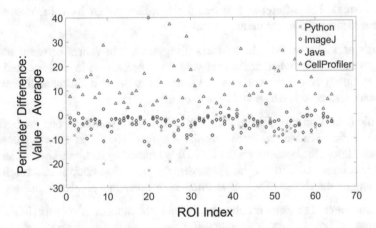

Fig. 3.14 Perimeter feature differences over multiple regions of interests (ROIs). The unit is image pixel

larger than 1 % across 64 test cells between the tools that agree and the ones that disagree. The "Absent" column is used to denote which tools do not have an implementation of a given feature.

Figure 3.14 illustrates the perimeter differences D_{ji} between its feature value V_{ji} and the average m_i of all three computed perimeter values per region of interest (i.e., cell segment). The feature difference follows the formula below:

$$D_{ji} = \left(V_{ji} - m_i \right) \tag{3.4}$$

$$m_i = \frac{1}{3} \sum_{j=1}^{3} V_{ji}$$

where i = 1,...,n; j = 1, 2, 3, j is the software index, and n is the number of ROIs. The perimeter values range between 44.4 and 542.5 pixels in the set of 64 ROIs (cells).

3.3.3 Discussion

Sources of Feature Variations
Next, we summarize our analysis of the sources of feature variations for the features listed in Table 3.1.

Perimeter and Circularity: The perimeter variability comes from the fact that algorithmic implementations differ in counting interior or exterior pixels, use four or eight connectivity of pixels, and might interpolate between the boundary points. Circularity is inversely proportional to perimeter squared.

Solidity: The same definition of solidity is used by Python and ImageJ (Area/Convex Area). The difference between these values comes from the convex area differences since the implementations vary.

Kurtosis and Skewness: The kurtosis disagreement in values between software packages depends on whether the excess kurtosis or kurtosis is implemented (fixed offset by three). Similarly, one must be aware of multiple definitions of skewness, for instance, sample versus population skewness.

Centroid (and Bounding Box): The centroid and the bounding box are both subject to the choice of the reference coordinate system (+col ~ x; +row ~ y or +row~ −y). In addition, the bounding box of a ROI is defined by its upper left corner coordinate and its width and height. However, the bounding coordinates might vary depending on the choice of values as integers or floats in a pixelated image.

Euler number: The Euler number definition is the number of objects (ROI) minus the number of holes. The value might differ depending on the assumptions about the number of ROIs (Python assumes to be one).

Histogram bins for intensities represented by more than 8 bits per pixel: ImageJ uses the max value plus one as the upper value of the last bin. It assumes that the lower value of the first bin is always zero.

Python and its NumPy library provide two definitions. B = histogram(X, N) uses N equally spaced bins within the appropriate range for the given image data type. The returned image B has no more than N discrete levels. B = histogram(X, edges) sorts X into bins with the bin edges specified by the vector, edges. Each bin includes the left edge but does not include the right edge. The last bin is an exception since it includes both edges. Python adjusts the automatic bin size selection to the input image class.

Orientation: The orientation is the angle between the major axis of a given ROI and the x-axis. It can be computed using two mathematical formulas: (1) $\theta = \text{atan}\left(\dfrac{V_y}{V_x}\right)$ where atan is the arctangent function and V_x and V_y are the x and y decompositions of the major axis of the ROI; (2) $\theta = \dfrac{1}{2} atan2\left(\dfrac{2I_{xy}}{I_{xx} - I_{yy}}\right)$ where I_{xx} and I_{yy} are the second moment of area along the x and y axes and I_{xy} is the product moment of area. These two formulas are equivalent if the first one is computed in the range of $[-\pi/2, \pi/2]$ using atan and the second one in the range of $[-\pi, \pi]$ using atan2. The variations are observed if different value ranges or angular units would be reported by selected software packages. Range can be either $[-\pi/2, \pi/2]$ or $[-\pi, \pi]$, and the unit can be either radian or degree. In addition, the sign of the output angle depends on the coordinate system (image coordinate or graph coordinate system with clockwise or counterclockwise axes).

Accuracy of Image Feature Implementations
After identifying feature variability among software packages and analyzing the sources of such variability, one could ask about the accuracy of image features. In other words, the task is to find a computed value that is the most accurate with respect to a ground truth value.

Accuracy analysis is based on two key components:

1. Generation of synthetic images and their corresponding reference feature values
2. A metric to compute the error between reference and computed values

For the ground truth, we use mathematical definitions of analog shapes and assume that synthetic digital images are very close representations of analog shapes. For the accuracy evaluations, we leverage the same software libraries as before but sub-select two image features for which we could generate reference feature values. Given the fact that we know the reference value, we could compute normalized relative errors E_i^r per ROI with respect to the reference feature value R_i (in comparison to the minimum value used in Eq. (3.5)).

$$E_i^r = |V_i - R_i| / R_i \tag{3.5}$$

where V_i is the measured feature value and i is the index of a ROI (image segment).

We report accuracy analyses for two image features including (a) major and minor axes of an ellipse and (b) a circumference.

Major and minor axis length: During the feature variability evaluation, we detected minor differences that were below the 1 % threshold on feature error. To test major and minor axis lengths, we created a set of 55 ellipse images with multiple values for major and minor axis length that ranges between 10 and 370. Figure 3.15 shows the normalized relative error E_i of major axis length computed according to Eq. (3.3) for ROIs in the 55 simulated images. It was observed that all three implementations had an error larger than 0.1 % when the ellipse shape was flat or the ellipse area was small. All implementations demonstrated the same dependency of feature error on ellipse shape/area.

Circumference: We created 23 synthetic binary images of a circle with radius ranging between $r \in [2, 222]$ pixels. The circle generation is done on images with size (500, 500) pixels using the following formula:

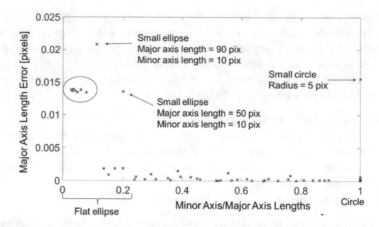

Fig. 3.15 Major axis length normalized relative error as a function of minor/major axis length for 55 synthetic ellipses

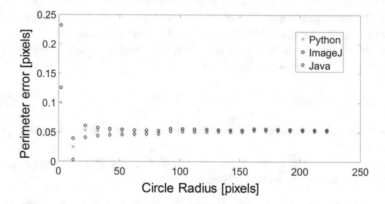

Fig. 3.16 Normalized circumference error vs circle radius

$$if\left(x-x_c\right)^2+\left(y-y_c\right)^2\le r^2\ then\ p\left(x,y\right)=1 \qquad (3.6)$$

where $x_c=249$ and $y_c=249$ are the coordinate of the circle centroid and r is the circle radius. The circumference reference value was set to $2\pi r$. Figure 3.16 displays the normalized error E_i computed as a function of circle radius according to Eq. (3.3).

The design of a synthetic image generator plays a significant role in representing the theoretical value and is always limited by the integer image lattice. For example, the results in Figs. 3.15 and 3.16 would be different if we placed the center of each ellipse at the lattice intersection as opposed in the middle of a square pixel (offset is

0.5). While the tools would still agree among each other, there will be bias between the computed major axis length or perimeter values and the reference value. This analysis shows that it is not accurate to analyze datasets that contain small ROIs.

3.4 Summary

This chapter presented the use of WIPP in three simple use cases. The use cases differed primarily by their input data, [x, y] versus [x, y, time], their types of static versus dynamic measurements extracted from very large FOV images, and their focus on obtaining numerical measurement versus measurement variability. Discussions provided additional insights that can be gained from the numerical results, such as the distributions of cell areas (use case 1), the identification of colonies that do not grow exponentially through time (use case 2), and accuracy of image features (use case 3).

Once cell or cell colony features are extracted from big image collections, the file sizes are reduced. The smaller file size implies that users can download tabulated results and continue analyses in other available tools. Interested readers can find detailed descriptions of more complex use cases online.[1]

References

1. van der Walt, S., et al.: scikit-image: image processing in Python. Peer. J. **2**, e453 (2014)
2. Schindelin, J., et al.: Fiji: an open-source platform for biological-image analysis. Nat. Methods. **9**(7), 676–682 (2012)
3. I. NIST.: "Image Features," web page, 2016. [Online]. Available: https://isg.nist.gov/deepzoomweb/stemcellfeatures. [Accessed: 31-Mar-2016]

---

[1] https://isg.nist.gov/deepzoomweb/software/wipp

Chapter 4
Components of Web Image Processing Pipeline

4.1 Mapping Functionality to Information Technologies

In this section, we map the functionality of WIPP to information technology components based on a simple WIPP usage scenario. Next, we classify the components into categories and present a summary of components used in WIPP.

Functionality and usage scenario of a web image processing pipeline
Let us consider a simple scientific discovery scenario to introduce the functionalities needed for WIPP. In this scenario, a user wants to use WIPP to visualize and analyze microscopy images of cells. The images are acquired as a grid of small fields of view that need to be assembled into one large field of view. The assembled large field of view is then displayed in the web browser with the ability to zoom in and out of the large field of view to visually explore the data. After visual inspection, the user performs image processing and analysis to extract measurements from the acquired images, for example, by segmenting cells and extracting their average intensity and area measurements.

Figure 4.1 illustrates functionalities of a web system that can support such a scientific discovery scenario. The left and right blue clouds refer to a user interacting with the data via a client (web browser). The middle part shows a server that hosts the data and metadata. The left green cloud refers to a computational cloud environment where the off-line processing computations are executed.

The scientific discovery scenario shown in Fig. 4.1 requires the web system to deliver the following general functionalities:

1. *Browser upload*: A user uploads a collection of image tiles in TIFF file format from a web browser (the blue cloud in Fig. 4.1, left). The server stores the uploaded data, extracts metadata from the TIFF files, and converts them into a standard format. The standard file format is the Open Microscopy Environment (OME) TIFF format which contains an OME-XML metadata block in the file header and the OME data model. The store-extract-convert functions require an underlying database for

© Springer International Publishing AG 2018
P. Bajcsy et al., *Web Microanalysis of Big Image Data*,
https://doi.org/10.1007/978-3-319-63360-2_4

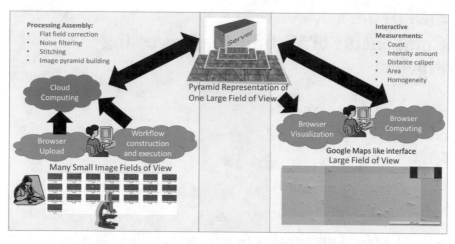

Fig. 4.1 Functionalities of a web image processing pipeline from a user perspective

storing data and metadata (e.g., MongoDB is used in WIPP), web user interface (UI) for uploading images, algorithms for executing the OME-TIFF conversion, and software for communicating between the database, user-driven upload, and computations (e.g., Spring Framework used in WIPP).

2. *Workflow construction*: A user specifies a set of tasks to execute, for example, flat field correction, noise filtering, stitching computations, and image pyramid building. These computations run custom-developed algorithms or existing algorithms in the widely used libraries (e.g., ImageJ/Fiji). The example sequence of four computations can be represented as a very simple workflow. Construction of such a workflow can be done via a web UI as shown in the blue cloud in Fig. 4.1 (left) and denoted as workflow construction and execution.

3. *Workflow execution*: The workflow is executed by sending the computational workflow description to the server to be parsed and executed by a workflow management system (e.g., Pegasus used in WIPP). The workflow execution faces several real-life challenges. For instance, a typical web system (a) serves multiple users, (b) consists of many computational hardware resources, and (c) utilizes many network drives with computers connected via local or wide area networks. This computational environment is denoted as cloud computing in Fig. 4.1 (green cloud). WIPP has a workload management system for compute-intensive jobs that addresses the challenges of execution time versus workload, reliability and failure recovery, and utilization versus cost. The main responsibilities of workflow and workload management systems are to distribute computations and data across available resources, schedule their job execution, maximize the utilization of each computational resource, monitor the progress of executions, and reassign executions and data distributions when failures occur.

4. *Browser visualization*: A user views the processed image stored on a server in a web browser (the blue cloud in Fig. 4.1, right). The image rendering is achieved

by using the HTML communication protocol between the browser and the server, and by fetching the pyramid tiles of interest. The rendering takes place in the web browser and is based on an open-source project called OpenSeadragon written in JavaScript. It is the HTML protocol and JavaScript that facilitate the pan and zoom-in/zoom-out functionalities when viewing large images. In addition, if a user needs to extract additional information from the images and render web pages with dynamic content (e.g., display pixel intensities while hovering a mouse of a pixel location or selecting image overlays), then more complex JavaScript libraries and frameworks can be used to build web application (e.g., AngularJS in WIPP).

5. *Browser computing*: A user enhances images and computes interactively measurements in a web browser (the blue cloud in Fig. 4.1, right). These computations are executed in the web browser by using JavaScript libraries for image enhancement (e.g., CamanJS used in WIPP), thresholding, segmentation, and counting. In addition, all measurements are supported by sub-setting and downloading options over terabyte-sized gigapixel frame images using the Web Deep Zoom toolkit.

Information technologies for a web image processing pipeline
In order to support the previous scenario, one can use client-server architectures and leverage multiple existing technologies to make users more powerful and productive when they conduct image-based research. We list several categories of information technologies found in client-server systems:

- Storage of data and metadata
- Algorithmic automation
- Algorithmic execution management in a computer cloud
- Visualization and computations in a web browser

Specific information technologies used in client-server system like WIPP are summarized in Table 4.1.

We will briefly describe each information technology in the remainder of this chapter while devoting Chap. 5 to automation of the image processing via algorithmic designs and Chap. 6 to algorithmic implementations that combine software and hardware. A reader will gain high-level understanding of which technologies can be used for delivering the functionality of systems like WIPP. For more in-depth understanding of each technology, readers are referred to additional published literature.

4.2   The Basics of Client-Server Architecture

In this section, we introduce readers to the roles of clients and servers in WIPP which is based on client-server architecture. A reader can find more information in textbooks about client-server architectures, for instance, [13–15].

Table 4.1 A list of open-source technologies used in building the web-based system and their functionalities

User-driven functionality in Fig. 4.1	Purpose	Technology name	Reference	Category
Browser upload	Database for storing information about image collections (such as list of images and metadata files, size, and provenance information)	MongoDB	[1]	Storage of data and metadata
Browser upload	Image metadata representation	Bioformats library from Open Microscopy Environment (OME)	[2]	Storage of data and metadata
Browser upload	Database access and web service implementations	Spring Framework	[3]	Storage of data and metadata
Algorithmic configuration	Image processing on a server	ImageJ and Java Advanced Imaging (JAI) libraries	[4], 5]	Algorithmic automation
Algorithmic configuration	Custom image processing on a server	Microscopy Image Stitching Tool (MIST)	[6]	Algorithmic automation
Workflow execution	Workflow management system Database for storing provenance information about computational jobs configurations	Pegasus	[7]	Algorithmic execution management in a computer cloud/ storage of data and metadata
Workflow execution	Workload management system for compute-intensive jobs	HTCondor	[8]	Algorithmic execution management in a computer cloud
Browser visualization	Building single page web applications (SPA)	AngularJS framework	[9]	Visualization and computations in a web browser
Browser visualization	Viewing with the pan-zoom functionality	OpenSeadragon JavaScript library	[10]	Visualization and computations in a web browser
Browser computation	Image filtering in a browser	CamanJS JavaScript library	[11]	Visualization and computations in a web browser
Browser computation	Extended functionality of OpenSeadragon for collaborations and measurements	Web Deep Zoom Toolkit (JavaScript library)	[12]	Visualization and computations in a web browser

4.2.1   The Role of Each Technology in the Client-Server Architecture

Typical roles of clients and servers
In the first chapter (see Sect. 1.2, "What does web image processing pipeline consist of?"), we introduced the concept of a client and a server. In a general client-server architecture, each computer or a process on the network serves a role of either a client or a server or both.

The following is an example of a client-server architecture: A company stores its products information in a database, which is used by a web server application to serve the data. Customers use the company's website or mobile application to access the product information. In this case, the database acts as a server for the web application, which acts as both a client (to the database server) and a server (for the website and mobile application clients).

The role of a client is to communicate with servers to request resources from the server (e.g., images or processing time). A client can be a local application used to render the received images or post-process requested data.

A server is typically a machine that is more powerful than a client in terms of computational and storage capabilities. It could also be the master of multiple machines (denoted as workers or agents or slaves). The role of a server is to respond to all requests from its clients, serve data, perform requested computations, or make sure that the computations are executed by its workers. Computers on the network that are serving as masters or workers are also frequently denoted as computing nodes.

Client-server architecture versus peer-to-peer architecture
Client-server and peer-to-peer architectures are two popular network architectures in distributed computing ([15], see Chap. 17). In the context of a "big image processing pipeline," a client-server architecture is preferred over a peer-to-peer architecture where each computer node has equivalent responsibilities. The reasons for the client-server preference lie in centralized security databases to control access, higher stability of the entire system than in a peer-to-peer architecture, and easier scalability to support many users because of its extensible, component-based platform and deployment. The peer-to-peer networking environment is cheaper but suitable only for low-cost small-size offices. The client-server architectures can consist of multiple tiers. Each tier corresponds to handling the user interface, database, and/ or application logic (two- or three-tier architectures).

The roles of clients and servers in a web image processing pipeline
Based on the categorization in Table 4.1, we associate the technologies with a general client-server architecture used in the design of WIPP shown in Fig. 4.2. The general architecture includes two types of clients, such as web browser clients and computational worker clients (see Fig. 4.2). One could also view multiple network area data storage devices as clients. The server has the role of a master and contains blocks of code for communicating with all types of clients. Browser upload in Table 4.1 (brown) maps to "data storage," "data access back end," and "web front

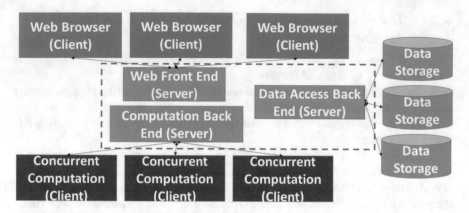

Fig. 4.2 A web image processing pipeline decomposed into server components (blue), data storage (brown), computational clients communicating with a server (red), and web browser clients communicating with a server

end" in Fig. 4.2. Workflow construction and execution in Table 4.1 are deployed as "concurrent computation clients" and "computation back end" in Fig. 4.2. Finally, browser visualization and browser computing in Table 4.1 maps to "web browser" in Fig. 4.2.

4.3   The Basics of Web Servers and Browsers

Web servers
To run client-server web applications, there must be a program running on a server that accesses web page files and that processes and delivers them to clients (i.e., web browsers). This program uses the Hypertext Transfer Protocol (HTTP) for communication between the server and the client and is denoted as the web server or the HTTP server. Web servers (1) receive a request from a client, (2) map the path of a Uniform Resource Locator (URL) into a file residing on a server for static requests or into a program name for dynamic requests, and (3) send the web page content to the client. The programs supporting dynamic requests reside on the server and are written in server-side scripting languages. The scripting languages allowing dynamical generation of web content include (a) PHP (for the recursive acronym PHP: Hypertext Preprocessor[1]), (b) Active Server Pages (ASP), and Java Server Pages (JSP).

Examples of web servers
There are many implementations of web servers and many middleware packages that include web servers. Among the implementations, we should mention open-source solutions, such as Apache HTTP server, Apache Tomcat server (Java), Gunicorn (Python Web Server Gateway Interface (WSGI)[1]), Waitress WSGI Server

---
[1] http://php.net/manual/en/faq.general.php

(Python), and uWSGI (C). The web servers are packaged in software products, such as Red Hat JBoss Web Server (Java), Spring Boot (Java), Django's Development Server (Python), or Tornado (Python). Some commercial products also embed open-source web servers and advance their features, for instance, the IBM® HTTP Server that is based on the Apache HTTP Server and additional IBM enhancements. When designing web systems like WIPP, one can choose a web server implementation that makes it compatible with the majority of all other components in terms of the programming language. Specifically, WIPP is using a Java-based Apache Tomcat web server since many other web components have been written in Java.

Standards for web servers

Features of web servers range from the basic HTTP request handling to advanced "on-the-fly" content generation. The features of Apache Tomcat server have to remain in sync with the current HTTP standards defined by the Internet Engineering Task Force (IETF®). The IETF is concerned with the evolution of the Internet architecture and with the compatibility of the server-side content and client-side rendering of the retrieved content. The IETF mission is to make sure that a large, heterogeneous collection of interconnected systems can be used for communication of many different content types.

Web Browsers

Web browsers are client-side software application for retrieving, presenting, and traversing information resources on the World Wide Web [16]. The retrieval location is defined by the URL. The URL starts with a scheme such as http:, https:, ftp:, or file: which indicates how to address the resource and can refer to the type of communication protocol used for communicating with local or remote resources. The rendering of the content is executed by the browser's software that transforms HTML markup language to a display according to the received content (text, images, audio, video, XML files, formatting instructions in CSS files, and dynamic content definition in JavaScript language).

The browser's software is running on the client's computer and devices. The HTML markup language can be written as either fat clients or thin clients (or hybrid clients). The key difference between fat and thin clients is that the fat clients rely on storing information on the client's local storage and utilizing client's CPU resources for computations other than just web page rendering. A thin client primarily renders web pages received from a web server which performs all computations. A hybrid client can be built that would store information on the server's computer but utilize client's CPU resources. In WIPP, we are using this type of hybrid client for any on-demand image filtering in the browser.

Examples of web browsers

New and updated web browser implementations[2] have been released every year since the introduction of the World Wide Web in 1991. The updates are important to incorporate changes that allow rendering of continuously evolving representations

---

[2] https://en.wikipedia.org/wiki/List_of_web_browsers

of web content. For example, the following new major versions of notable browsers were released in 2017: Google Chrome, Microsoft Edge, Mozilla Firefox, Opera, and Opera Neon. These browsers differ in terms of their web content support. In order to render the content using any web browsers, web developers insert code for detecting the browser type and adjusting the content rendering according to the web browser type.

In addition to the variety of web browser implementations, there is a variety of hardware limitations for rendering web pages (i.e., a variety of displays on mobile devices). To guarantee correct rendering of HTML web pages, a web developer has to take into consideration browser type and its release version and hardware device and its display limitations. To address the heterogeneity of software and hardware environments, the World Wide Web Consortium has been developing a hybrid of XML and HTML markup language for web pages called XHTML (Extensible Hypertext Markup Language).[3] The use of XHTML would make web pages more likely to interoperate within and among various environments.

The web browser code in the WIPP is designed in HTML and is assumed to execute in browsers on regular computers. WIPP does not have a special support for mobile devices and has been primarily tested using Google Chrome, Mozilla Firefox, and Apple Safari. Developers can redesign portions of WIPP using XHTML and insert code for adjusting the content rendering according to the web browser type.

4.4   The Basics of Communication Protocols in Client-Server Architectures

The general client-server architecture presented in Fig. 4.2 is expanded in Fig. 4.3 to include not only the technology names but also the communication protocols. We already mentioned HTTP as the communication protocol between web browsers and web servers. In the case of WIPP, we are using the Java Spring Framework and the Apache Tomcat web server (Spring Boot) to handle all HTTP or secure HTTPS communications. On the server side, the communication between Java Spring Framework and applications uses HTTP to retrieve status of computations. The computational job management is facilitated by the Pegasus scientific workflow management system. Pegasus has modules for communicating using Transmission Control Protocol (TCP) and Message Passing Interface (MPI) protocols between the master and worker computational nodes. Finally, the communication between a file system (disk) and the applications is enabled via Network File System (NFS). We briefly introduce these communication protocols next and refer a reader to existing books about HTTP [17], TCP [18], MPI [19], and NFS [20] for more details.

---

[3] https://www.w3schools.com/html/html_xhtml.asp, https://www.w3.org/TR/xhtml1/

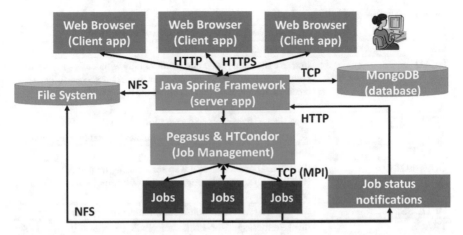

Fig. 4.3 Specific client-server architecture of the deployed web image processing pipeline (focus on server side)

4.4.1   Client-Server Communication Using Hypertext Transfer Protocol

Hypertext Transfer Protocol (HTTP)
The communication between the web browser and the web server uses the Hypertext Transfer Protocol (HTTP), a request-response protocol. This communication takes place after typing a web address in the browser to access a web server or a web service. The web address is expressed as a Uniform Resource Locator (URL).

HTTP methods
The HTTP protocol (version 1.1) contains implementations of eight methods: GET, HEAD, POST, PUT, DELETE, CONNECT, OPTIONS, and TRACE. The request for comments (RFC 5789) adds the ninth method: PATCH. Table 4.2 summarizes these methods. The methods are also classified as *safe* and *idempotent* operations. Safe methods only retrieve information, for example, GET or HEAD. Idempotent operations are those that when executed many times they have the same side effect as when executed only once, for instance, GET, HEAD, PUT, OPTIONS, and DELETE. One should note that while the Representational State Transfer (REST) Application Programming Interface (API) remains just as software architectural style, its implementation with the HTTP communication protocol has become a standard.

Secure Hypertext Transfer Protocol (HTTPS)
Figure 4.3 denotes the web browser and server communication as HTTPS which refers to a secure HTTP. The secure communication is achieved by encrypting messages using Public Key Infrastructure (PKI) and by adding Secure Sockets Layer (SSL) and Transport Layer Security (TLS) encryption layer on top of the HTTP protocol. The secure HTTP communication consists of three steps:

1. *Step 1: Hello messages* – A web browser sends a Hello message to the web server with the information about cipher suites and SSL version that the web

Table 4.2 Summary of HTTP methods

HTTP (version 1.1) method	Method description from a web browser viewpoint
GET	Retrieve data from web server
HEAD	Retrieve the status code and header fields without any message body. Status code indicates whether the server successfully received, understood, and accepted the request. Header fields contain content type and URI
POST	Send data to the server (e.g., web form inputs)
PUT	Replace current content with the uploaded content on the server
DELETE	Remove current content at a URI
CONNECT	Establish a connection to a web server using tunneling (i.e., via HTTP proxy server that forwards the web browser request to the desired server)
OPTIONS	Provide the request-response communication options
TRACE	Invoke a message loop-back after sending a request in order to see what is being received at the server side
PATCH	Applies partial modifications to the resource

browser supports. The cipher suites define the details of encryption and decryption algorithms (an encryption key exchange algorithm used during authentication, a bulk encryption algorithm to encrypt the message, a message authentication code (MAC) algorithm to create a cryptographic hash, and a pseudorandom function (PRF) used by the MAC algorithm's hash function). The server responds with a Hello message containing similar information and a decision which cipher suite and version of SSL will be used.

2. *Step 2: Certificate exchange* – The web server sends its SSL certificate to the web browser. The SSL certification includes information about its owner's identity, and the digital signature of an entity that has verified the certificate's contents are correct. The web browser checks whether to trust the web server or not based on a presented SSL certificate and its verification. The web server is also allowed to require a SSL certificate from a web browser for exchanging sensitive information.

3. *Step 3: Key exchange* – Following the agreed SSL version and a shared public key of the web server in the SSL certificate, a web browser generates a random key, encrypts it with the web server's public key, and sends it to the web server. The web server decrypts the message by using the corresponding private key stored on the web server. This decrypted key can be then used by the web server for secure communication with the web browser. This is called a symmetric algorithm for HTTPS communication.

4.4.2 Transmission Control Protocol (TCP)

A web server is typically a computer connected to other computational and storage resources via local area network (LAN). WIPP is using the Transmission Control Protocol (TCP) to communicate between the Java Spring Framework (web server application) and the computing nodes (i.e., computers performing computation denoted as "Jobs" in Fig. 4.3). This protocol is frequently referred to as TCP/IP since it originated from the Internet Protocol (IP). TCP/IP guarantees reliable, error-checked delivery of data packets in the same order in which they were sent. If data packets are lost or duplicated or delivered out of order due to a network congestion or a lack of network reliability, TCP requests retransmission of lost data packets, removes duplicates, and sorts out-of-order data. If the data packets are undelivered, then the sender is notified of this failure.

Use of TCP

TCP has been used by many Internet applications, such as File Transfer Protocol (FTP), peer-to-peer file sharing, on-demand streaming media applications of pre-recorded video, email, and the World Wide Web (WWW). Following Fig. 4.3 as our simple illustration, a Java Spring Framework (web server) sends a file to a client machine managing computational jobs via the TCP software layer. The TCP software divides the sequence of file bytes into segments and forwards them individually to the Internet Protocol (IP) software layer. The IP software adds a header with the destination IP address of the client machine to the TCP segment and creates an IP packet. The TCP software at the client machine receives the IP packets, recreates the individual TCP segments, checks that they are error-free and correctly ordered, and then sends them to an application that manages computational jobs.

4.4.3   Message Passing Interface

Once a job execution message came via HTTPS from the web browser to the web server (Java Spring Framework) and via TCP to the machine managing computational jobs, there is a need to communicate with multiple computing nodes (red boxes with the label "Jobs" in Fig. 4.3). The purpose of this communication is to execute multiple jobs in parallel to speed up the computation and leverage all computing nodes.

Message Passing Interface (MPI) has been designed to write software for Single Process, Multiple Data (SPMD) parallel applications. SPMD applications are typically run on distributed memory computer architectures which consist of a collection of independent computers, called nodes. Each computer node starts its own program/process and communicates with other nodes by sending and receiving messages. MPI can be viewed as a communication specification for point-to-point and collective (group) communication. The MPI de facto standard specification defines the syntax and semantics of library routines for writing portable message passing programs.

MPI implementations

Using the message passing parallel programming model, data can be moved from the address space of one computational process to that of another process through cooperative operations on each process. There are 430 routines defined in MPI, version 3. A software developer has to identify parallelism and implement parallel algorithms using MPI routines. The MPI library routines have been designed to utilize a variety of computer node architectures (see examples in Fig. 4.4). Most MPI implementations have application programming interface (API) to routines that are

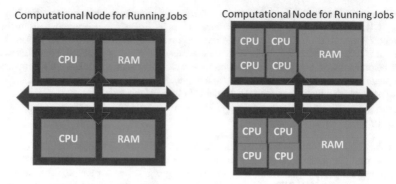

Fig. 4.4 Illustration of computational node architectures creating hybrid distributed memory/ shared memory systems

directly callable from C, C++, Fortran, and any language able to interface with such libraries (e.g., C#, Java, or Python). MPI has been implemented for almost every distributed memory architecture and is optimized for the hardware on which it runs.

Use of MPI for cluster computing

In WIPP, we use the implementation of a scientific workflow management system called Pegasus. Pegasus contains a component denoted as pegasus-mpi-cluster[4] that leverages MPI (although not currently used directly by WIPP). The applications following the MPI interface are represented by Directed Acyclic Graphs (DAGs) where each node in the graph is a computational task/job and each edge defines the execution order dependency between the tasks. Each computational task needs to be run from a command line with some optional arguments. The dependencies describe the data flow in which the output files produced by one task are needed as inputs for another. A Pegasus-mpi-cluster job consists of a single master process and several worker processes that receive workflow tasks for execution and return the results to the master via the MPI messages.

4.4.4 Network File System

We are using the Network File System (NFS) to access files over LAN the same way as one would access the files on a local disk. NFS follows an open standard and uses remote procedure call (RPC) system (i.e., the Open Network Computing Remote Procedure Call (ONC RPC) system). The ONC RPC system:

(a) Serializes data using the External Data Representation (XDR)
(b) Transfers the data in XDR using either TCP or UDP (User Datagram Protocol)
(c) Allows access to RPC services on a remote machine via a port mapper that listens for queries on a well- known port 111 using TCP or UDP protocol

---

[4] https://pegasus.isi.edu/documentation/cli-pegasus-mpi-cluster.php

Use of NFS In the Unix platform WIPP configuration, the TB-sized image collections and their pyramid representations can be stored on a remote machine (i.e., network-attached storage (NAS)). Thus, the NAS machine becomes the NFS server, and the web server becomes a client requesting access to data. Once the NFS server is installed and configured for data access to clients, the client machine requests access to the NFS server by issuing a mount command. Because of mount, the client can interact (read, write, view) with the mounted file system of the NFS server according to the NFS system permissions.

4.5 Designing Interactive User Interfaces in Web Browsers

General best practices and guidelines for building web applications can be found in many books on the market, for instance, [21–23]. A reader can gain in-depth knowledge about the key components of web application design, such as HTML, Cascading Style Sheet (CSS), JavaScript, and Document Object Model (DOM).

WIPP leverages many web components that provide interactive user interfaces to very large images and their associated computations. The interactive user interfaces in WIPP follow a Model-View-Controller design pattern and leverage several existing JavaScript libraries and frameworks. To achieve interactive user interfaces, WIPP depends on existing JavaScript frameworks and libraries, such as AngularJS, Data-Driven Documents (D3.js), JQuery, OpenSeadragon, and CamanJ. Each of these JavaScript libraries has web pages with documentation for the interested reader. In this section, we introduce the basics of a Model-View-Controller design pattern for writing the code running in web browsers and then focus on the use of AngularJS for creating interactive user interfaces in WIPP.

4.5.1 Model-View-Controller Design Pattern

The Model-View-Controller (MVC) pattern is one of the most used design patterns for user interface design and web programming. It is described in detail in [24]. Figure 4.5 shows the overview of the Model-View-Controller (MVC) design pattern. The View part of the code is the user interface portion that is made typically with HTML, CSS, and JavaScript code. For example, when a user selects an option from a drop-down menu, the action is propagated to the Controller. The Controller communicates with the Model to display a submenu of the main menu or to mark the selected menu as chosen. The Model part of the code stores the application specific data, for example, the options of a main menu and its submenus or the menu choices selected by a user. As soon as the Controller receives the notification about the submenu entries from the Model, it passes the information to update the View. The View can render the content of submenus or show a check mark next to a selected main menu.

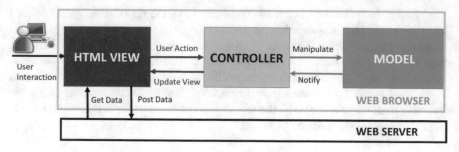

Fig. 4.5 Model-View-Controller (MVC) design pattern for building JavaScript code

4.5.2 AngularJS for Building Interactive User Interfaces

When writing the code that follows the MVC design pattern, the developer's goal is to make the HTML web pages interactive (dynamic HTML web pages). Dynamic web pages perform rendering of content in response to user's interactions. The interactive user experience is achieved by a combination of HTML and JavaScript. To accommodate multiple interfaces in interactive views, a Cascading Style Sheet (CSS) file is added to describe how the HTML elements are to be displayed and the layout of the rendered elements in each view. Dynamic HTML is a combination of HTML, a client-side JavaScript scripting language, a presentation description in CSS files, and the Document Object Model (DOM).

To design interactive user interfaces for large image visualization and processing in WIPP, we used AngularJS framework as it simplifies the developer's effort to manipulate the HTML content. We will briefly highlight the key advantages of AngularJS JavaScript framework, such as data binding between DOM and HTML page, directives, and deep linking.

Introduction to Document Object Model (DOM)

The Document Object Model (DOM) is an application programming interface (API) for HTML documents. DOM API defines methods for accessing and manipulating a tree structure representation of the HTML content.[4] Each tree node is an object representing a part of the HTML document. By using JavaScript, the objects can be manipulated programmatically and object changes in the HTML document can be rendered. For example, HTMLDocument interface allows to access the root of the HTML hierarchy and to hold the entire content. The entire content consists of:

- Strings (title, referrer, domain, URL, cookies)
- HTMLElement (body)
- HTMLCollections (images, applets, links, forms, anchors)
- Several basic methods (open, close, write, getElementsByName)

HTMLCollection interface provides access to a list of nodes by either ordinal index or the node's name or id attributes. HTMLElement interface is for accessing basic HTML components. In general, the DOM API can be used in a wide variety of applications (not just web applications) and with any programming language. The DOM API standard5 has been developed by the World Wide Web Consortium (W3C).

Data binding between DOM and HTML page

AngularJS[5] is a JavaScript framework for building dynamic web applications using HTML, JavaScript, and CSS files. It eliminates DOM manipulation by binding data in the DOM representation with the view in HTML web pages. The term "binding" refers to the connection between a user interface (HTML web page) and the data model (values of web page variables in DOM). AngularJS enables a "two-way" data binding. If the user edits a value in an editable text HTML element, then the value is automatically updated in the DOM representation to reflect that change. If a value is modified in the model, then the view is automatically updated as well.

Directives to extend HTML syntax

In addition to data binding, AngularJS allows a developer to create functions associated with DOM elements (called directives) that can extend the HTML syntax. Directives are translated to HTML, CSS, and JavaScript when they are rendered in a browser. The advantage of directives lies in the improved expressiveness of HTML for domain-specific behavior. Using AngularJS directives, HTML tags can be associated with user-defined functions which lead to more customized web applications.

Deep linking

Another useful feature of AngularJS is deep linking. This term refers to the case when a web page does not contain a single web application but consists of multiple web applications. Since multiple web applications can have different states, bookmarking and back browser navigation become a problem. In the simplest case, deep linking can be understood as the ability to hyperlink to any piece of web content on a website (i.e., a paragraph on a web page). In the most complex case, it is the ability to restore a web page and of the all states of its web applications. AngularJS can manage the mapping between the current state of the page and the corresponding web application sub-templates. Depending on the progress status of a set of web applications (or the state of the web page), AngularJS would reload the sub-templates by setting up different URLs (hence the name "deep linking").

4.6   Large Image Visualization and Processing in Web Browsers

The introduction of web servers, web browsers, communication protocols, and interactive web user interfaces is the prerequisite to addressing challenges with big image data in WIPP. The initial functionalities of interest are viewing gigapixel images, interacting with the images in real time, and storing information derived

---

[5] https://angularjs.org/

from visual explorations. Given gigapixel images, this section will build on top of the previously introduced technologies and lay the grounds for:

1. Representation of large images
2. Large image visualization in web browsers
3. Image processing in web browsers

4.7   Representation of Large Images

It is envisioned that WIPP users will use regular monitors as display devices. The interplay between the gigapixel images stored on a hard disk and the monitor screens requires considerations about scalable representation, organization, and file formats for storage and retrieval.

Scalable representation

Large image visualizations must use a scalable data representation. The word "scalable" refers to the fact that a laptop screen size is much smaller than gigapixel images. Current sizes of liquid-crystal displays (LCD) vary between 12″ and 17″. They provide a user with an average of 1.37 viewable Megapixels (between 0.48 and 2.3 Megapixels – see Table 4.3). Thus, a scalable image representation for laptop screens must allow viewing an entire gigapixel image at coarse and fine spatial resolutions over viewable Megapixel screen sizes.

Image pyramid representation

To address the scalability, one can create versions of gigapixel images at multiple resolutions and then tile them spatially into image subregions that fit on a laptop screen. This representation is called an image pyramid since the image size is decreasing with the coarser resolution. The tiles at each resolution level are of a

Table 4.3 Laptop screen resolutions

Screen resolutions (in pixels)	Possible LCD sizes (diagonal length in cm and in inches inside of parenthesis)	Viewable Megapixels
800 × 600 (SVGA – standard)	30.48 (12)	0.48
1024 × 768 (XGA – standard)	30.48 (12), 33.78 (13.3), 35.56 (14), 38.1 (15)	0.79
1280 × 800 (WXGA – wide)	30.73 (12.1), 33.78 (13.3), 35.81 (14.1), 39.12 (15.4)	1.02
1440 × 900 (WXGA+ – wide)	35.56 (14)	1.30
1280 × 1024 (SXGA – standard)	35.56 (14), 38.1 (15), 39.88 (15.7)	1.31
1400 × 1050 (SXGA+ – standard)	30.73 (12.1), 35.56 (14), 38.1 (15)	1.47
1680 × 1050 (WSXGA+ – wide)	39.12 (15.4)	1.76
1600 × 1200 (UXGA – standard)	35.56 (14), 38.1 (15), 40.64 (16)	1.92
1920 × 1200 (WUXGA – wide)	39.12 (15.4), 43.18 (17)	2.30

According to http://www.geek.com/laptop-screen-size-resolution/

Fig. 4.7 Image pyramid representation

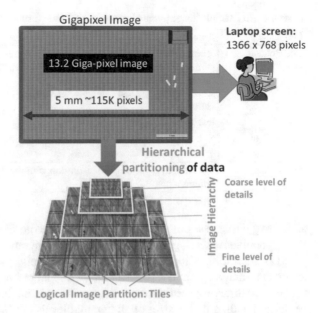

fixed size, and they correspond to logical image partitions (i.e., neighboring pixels create visually meaningful image content). Figure 4.7 illustrates the image pyramid representation for a large field of view image acquired by a microscope. The process of creating an image pyramid is called hierarchical partitioning. The process can be described as iterative filtering followed by subsampling each filtered image by a factor of two along each image dimension. One of the simplest methods is to iteratively replace every 2×2 pixels with their average.

Heterogeneity of image pyramid formats

Small images are acquired by cameras embedded in telescopes, satellite and aerial imaging instruments, cell phones, photographic devices, and microscopes. Large images are typically formed by stitching small images of adjacent fields of views. Image pyramid representations of large images have been built in several application domains including astronomy, Geospatial Information Systems (GIS), digital humanities, materials science, cell biology, medicine, and many more. Given the heterogeneity of imaging instruments and application domains, one can find a variety of image pyramid formats including Deep Zoom Images (DZI), Open Street Maps (OSM), Tiled Map Service (TMS), International Image Interoperability Framework (IIIF), and Legacy Image Pyramids (LIP).

DZI image pyramid format

For illustration purposes, we will describe the DZI format which is an XML specification maintained by Microsoft [25]. An image pyramid is stored as a set of folders containing image tiles and a metadata file describing the pyramid. Table 4.4 shows an example of the image pyramid storage for an image of size $(22\ 934 \times 21\ 056)$ pixels. The tile size is set to (258×258) pixels with one pixel overlap on each side. The overlap value indicates how many columns and rows on each side of a tile are

Table 4.4 An example image pyramid stored in a DZI file format

File name	test.dzi
File Content	<?xml version="1.0" encoding="utf-8"?> <Image TileSize="256" Overlap="1" Format="png" xmlns="http://schemas.microsoft.com/deepzoom/2009"> <Size Width="22,934" Height="21,056"/> </Image>
Folder	test_files
Folder Content	Sub-folders labeled from 0 to 15 (the number of pyramid levels)
Sub-Folder	15
Sub-Folder Content	Tile images at the original resolution with the file names <grid-row>_<grid-column>.png

added to guarantee seamless rendering. The image tile file format is specified to be PNG (portable network graphics). The number of pyramid levels is 15 and is computed as the ceiling of $\log2(\max\{\text{Width} = 22\,934, \text{Height} = 21\,056\})$. One should be aware that image pyramid representation of any large image will require about 4/3 times more storage space than the storage of the original image. The number 4/3 can be derived adding up the sizes of all pyramid levels $(1+1/4+1/16+....1/(2^{(2*\text{number of levels})}))$ ~(4/3).

Container file storage of image pyramids

Image pyramids are collections of files and folders that can reside on a file system or can be stored in one container file. For example, a BigDataViewer plugin[6] to Fiji uses the Hierarchical Definition Format version 5 (HDF5) for storing all files. HDF5 provides a container for chunked multidimensional arrays (image tiles). It keeps files together but has some overhead in reading and writing.[7] Another option is to use the multi-page Adobe Tagged Image File Format (TIFF), but the format is limited to 4 gigabytes. The ongoing effort to design a BigTIFF file format would allow to save files larger than 4 gigabytes but might still suffer from read/write overheads.

File formats of image tiles

Note that there is a variety of web browsers, and their image rendering support varies in terms of image file formats. The main image file formats that have been supported by most web browsers are jpg, png, and gif file formats. A detailed summary of image formats supported by the existing web browsers can be found online.[8]

Image pyramids with additional information

Every application domain has a need for displaying large images with additional information. For example, in cultural heritage applications, one can show

---

[6] http://imagej.net/BigDataViewer

[7] HDF chunking issues: https://support.hdfgroup.org/HDF5/doc/H5.user/Chunking.html

[8] https://en.wikipedia.org/wiki/Comparison_of_web_browsers#Image_format_support

high-resolution illustrations in books[9] or maps[10] with textural, graphical, and pictorial notes about history of depicted objects. In geospatial mapping applications, one would like to add geo-referenced overlays of information about vegetation, roads, hotels, restaurants, and so on, which has been implemented in web systems such as Google Maps and MapQuest. In biomedical applications, image pyramids can represent 2D+time or 3D large images by creating a series of pyramids. In materials science, image pyramids can be used for viewing large 2D images inside of another 2D large image (local high magnification view inside of global low magnification view) by creating an image pyramid inside another image pyramid (called sparse pyramids). All aforementioned cases leverage the image pyramid representation but require custom solutions to integrate additional information with the basic functionality of retrieving image tiles for visualization and further processing.

4.7.1 Large Image Visualization in Web Browsers

The scalable representation of gigapixel images as image pyramids requires software for rendering image pyramids in a browser and for retrieving image pixel information based on user inputs. WIPP leverages the OpenSeadragon library [10] and integrates it into Web Deep Zoom Toolkit (WDZT) described below.

Rendering image pyramids in a browser
An image pyramid representation is critical to accommodate the bandwidth limitation between a web server and a browser and the RAM limitations of computers running web browsers. If these two limitations can be overcome, then users can have interactive sessions while viewing gigapixel images in their browsers. During the interactive viewing, the browser has to use the HTML request-response protocol to fetch image tiles based on user's choice of zoom level (hierarchical partition level) and pan location (logical partition level) and then render the tiles in the web browser. These steps are accomplished by JavaScript running in the browser.

The original code for large 2D image visualization (also called Deep Zoom) was developed by Seadragon Software and later expanded by Microsoft Live Labs to a Silverlight product. Google has used similar zoom and pan concepts to support the delivery of Google Maps. The initial 2D support was extended to 3D for medical image volumes [26] and to other informative visualizations [27] with an open-source project behind many added functionalities. We will focus on OpenSeadragon JavaScript library[11] [10] that has originated from the Silverlight product and has been extended as an open-source project by many contributors. Our focus is on the extensions to OpenSeadragon called Web Deep Zoom Toolkit (WDZT). It is described in [28, 29] combined with the efficiency studies about building image pyramids from large 2D images [30–32].

---

[9] Google books: https://www.google.com/intl/en/googlebooks/about/index.html

[10] David Rumsey's map collection: http://www.davidrumsey.com/

[11] https://openseadragon.github.io/

JavaScript implementation for additional image analyses
Web Deep Zoom Toolkit (WDZT) is a JavaScript framework for analysis of deep zoom images on the web. It is accessible from the GitHub repository[12] and leverages OpenSeadragon library and several plugins for reporting pixel values (OpenSeadragonPixelColor), displaying an image scale bar (OpenSeadragonScalebar) and providing image filtering options (OpenSeadragonFiltering). The source code is divided into the:

(a) Core components
(b) Logic code that might be potentially reused by multiple modules
(c) Functional modules
(d) User interface widgets

For example, functional modules that extend OpenSeadragon are accessible from the left pane of WDZT (see Fig. 2.4 in Chap. 2) and are encapsulated in JavaScript files in the src/modules sub-directory. A new module can be developed by extending the module class and adding it to the src/modules directory. Debugging the new module can be accomplished by creating an HTML file and adding the new module to the HTML file.

4.7.2   Image Processing in Web Browsers

When designing systems for image processing, a developer faces a decision about where to execute computations (server versus browser) and how to orchestrate computations (synchronous versus asynchronous image processing). WIPP design incorporates these decisions to achieve interactivity of simple image processing operations, such as image filtering.

Server versus browser computation tradeoffs Interactive measurements in a web browser require executing computational operations beyond those for moving data between the web server and the web browser or the web browser cache and the screen buffer. For example, calculating a Euclidean distance between two mouse-clicked locations must be performed in the Arithmetic Logic Unit (ALU) inside of a computer's central processing unit (CPU). There is always a software design dilemma about how much computing should be done by the web browser as opposed to the web server. If all computations are done by the server, then users may experience large communication response delays. If all computations are done by the web browser, then users might experience a slow response due to an insufficient computing power to calculate the results. This situation is exacerbated when dealing with very large images and with computations requiring significant computer memory (RAM) and processing power (CPU).

---

[12] https://github.com/usnistgov/WebDeepZoomToolkit

Interactivity of simple image processing
Given the image pyramid representation, we assume that a computer running the
web browser has enough RAM and sufficient CPU power for simple image processing of image tiles rendered on a laptop screen. Simple image processing is useful for
image enhancements, thresholding, connectivity analysis, and counting of connected components. One such solution is the CamanJS[13] image enhancement library
written in JavaScript language. It enables to change contrast, hue, brightness, exposure, gamma, saturation, as well as colorize and clip images in a web browser. The
value of image enhancements is that a user can interactively choose parameters of
image processing operations by testing them on viewed image tiles of a pyramid.
The values obtained from browser-based image processing can serve as initial
parameter estimates for image processing operations that can be then applied to the
entire image by using a more powerful server computer.

Synchronous versus asynchronous image processing
To enable image processing over large images, one must integrate functionality of
CamanJS library with OpenSeadragon so that image processing can be applied to
image pyramids as opposed to single images. This is accomplished via the
OpenSeadragonFiltering plugin.[14] The processing can be performed asynchronously
or synchronously with the retrieved image tiles.

Asynchronous processing does not wait for the function call to return from the
server. A sequence of processing functions continues to be executed, and a "callback" function is executed when the asynchronous processing returns from the
server. Unlike asynchronous processing, synchronous processing waits for the function call to return before continuing with the execution. The advantage of synchronous processing is that the order of execution is known, but the trade-off is that the
current processing thread is blocked while waiting, which will cause the web
browser to "freeze" (i.e., be unresponsive) during that time. In practice, asynchronous processing is used for retrieving all pyramid tiles from the server and then for
applying a processor function to filter images. Thus, a confirmation callback is
returned from the server after all pyramid tiles are retrieved and filtered. The pyramid tile retrieval and filtering is asynchronous processing because the filtering can
be applied to pyramid tiles as the tiles arrive from the web browser. Asynchronous
execution is also faster because it can utilize multiple CPUs on the client side by
sending a filtering job with an image tile to a CPU as soon as the tile arrives. On the
other hand, image processing that operates over neighboring pixels needs multiple
image tiles, and therefore it must utilize synchronous processing or a custom image
tile retrieval before a computation is executed.

Asynchronous implementation of OpenSeadragonFiltering plugin
In the current implementation of the OpenSeadragonFiltering plugin, all image processing operations are executed asynchronously regardless of pixel-level manipulations or spatial kernel operations. Thus, spatial kernel operations are executed over

---

[13] http://camanjs.com/

[14] https://github.com/usnistgov/OpenSeadragonFiltering/blob/master/openseadragon-filtering.js

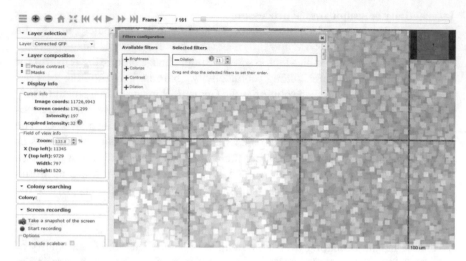

Fig. 4.8 Artifacts introduced by asynchronous retrieval and spatial filtering using the OpenSeadragonFiltering plugin

each tile independently (without any pre-defined order) and cannot handle image tile boundaries where multiple tiles would have to be synchronously retrieved. For instance, when Sobel edge detection or morphological dilation/erosion operations are applied, a user can see a border artifact especially when the image tile overlap is larger than the spatial kernel size. Figure 4.8 illustrated the artifacts for morphological dilation with the kernel size equal to 11. Irrespective of the artifacts, the value of image processing in a web browser lies in the interactivity of parameter estimation and visual inspection.

4.8   Managing Images, Pyramids, and Metadata

One of the key components of WIPP is the underlying database for storing information about images and computations applied to images. A reader interested in studying databases in the context of web systems is referred to textbooks, such as [33, 34]. In this section, we provide just a top-level overview of key concepts about database management systems, the two main types of databases, and a short description of MongoDB database [35] used in WIPP. The rest of the section is devoted to Java Spring Framework as one of web application frameworks for connecting the MongoDB with the other components in WIPP.

Purpose of databases

Storage and management of large amounts of data present challenges related to organization of data and retrieval of useful information. Databases provide a mechanism for storing collections of data in such a way that information of interest can be easily accessed, updated, and searched. Databases are used in many commercial web applications, such as online catalogs of products, users, and purchase orders.

In scientific applications, databases serve as catalogs of reference measurements, experimental data, metadata, and derived results for scientific discoveries.

Database management systems
To perform operations on data in a database, databases come with a collection of programs called a database management system (DBMS). DBMS becomes an interface between a database and all users or other software applications. The administration of databases with DBMS is achieved by executing four basic operations of persistent storage also called CRUD functions:

- C (Create): insertion of new data in the database
- R (Read): selection and retrieval of data using queries
- U (Update): modification of existing data
- D (Delete): suppression of existing data

Widely used DBMSs include (a) open-source solutions, such as MySQL,[15] PostgreSQL, and MongoDB, and (b) commercial solutions, for example, Oracle, Microsoft SQL Server, or IBM DB2.

Data object conversions
In object-oriented software applications, data objects must be converted from the database representation to the target programming language representation, and back. This conversion between database objects (records represented by rows in a table or documents in a collection) and objects in programming language (such as Java or Python objects) can be done in an automatic way using an Object-Relational Mapping (ORM) library. It can also be implemented using the native procedural language provided with the database in a Data Access Object (DAO) layer. The DAO layer provides an abstract interface to the rest of the application for interacting with the database.

Database classification
Databases can be divided in two categories:

1. Relational databases
2. Non-relational databases

The key differences lie in their representation of information. We outline the database representations in both categories and provide examples to illustrate the key differences.

4.8.1   Relational Databases

Representation
Relational databases are based on a tabular representation of the data where each table contains objects of the same type. Multiple tables are related according to common keys or concepts which is the basis for the term "relational database."

---

[15] https://www.mysql.com/

Table 4.5 Comparisons of requirements on primary and foreign keys

Requirements	Primary key	Foreign key
Duplicate values are allowed	No	Yes
Null values are allowed	No	Yes
Number of columns	One	One or more

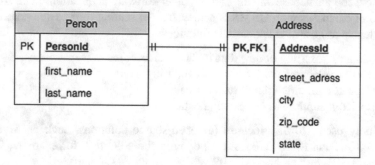

Fig. 4.9 Entity relation model

The tables can store primary and optional or non-primary information in separate tables and preserve the information relations.

Representation example
For example, a person is often represented by primary information such as his first and last name. Each person has also an address viewed as non-primary information. In this case, a table *Person* is storing people's first and last names, and another table *Address* will be storing address information such as street address, city, zip code, and state. The "relation" link between a person and his address is represented with a virtual key. The virtual key in a table can be primary or foreign. The keys must satisfy a set of requirements in terms of uniqueness as shown in Table 4.5. For the Person-Address example, the primary key is *PersonID* associated with each person in the *Person* table (Fig. 4.9). The foreign key in the *Address* table uses the values of the *PersonID* key to associate each address with a person. Note that one person can have multiple addresses, and therefore the foreign key column can contain duplicates of *PersonID* key values.

Relations between entities can be of three types:

1. One-to-one
2. One-to-many
3. Many-to-many

One-to-one relation: A one-to-one relation between two types of entity A and B implies that an entity A can only have a relation with one entity B, and vice versa. For instance, a country can only have one capital city, and a city can only be the capital of one country as shown in Fig. 4.10.

Fig. 4.10 One-to-one
entity relation diagram

Fig. 4.11 One-to-many
entity relation diagram

Fig. 4.12 Many-to-many
entity relation diagram

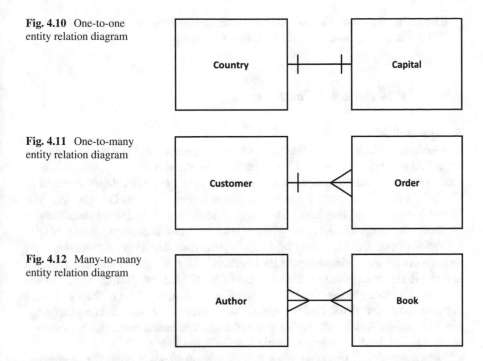

One-to-many relation: A one-to-many relation between two types of entity A and B means that an entity A can be linked to several entities B, but an entity B can only be linked to one entity A. For instance, a customer can place several orders on a commercial website, but each order will only be linked to exactly one customer as shown in Fig. 4.11.

A many-to-many relation between two types of entity A and B means that an entity A can be linked to several entities B, and vice versa. For instance, an author can write several books, and a book can be written by several authors as shown in Fig. 4.12.

Properties: Relational databases follow the ACID (Atomicity, Consistency, Isolation, and Durability) properties for their operations to assure consistency of the data in the case of simultaneous transactions:

- *Atomicity* ensures that a transaction with data in database is either completed or rolled back to the original state. In other words, there is no partial state of a transaction.
- *Consistency* requires that when a transaction is completed that it meets consistency checks. For example, moving money from an account A to another account B implies that the sum of A + B is the same before and after transaction completion.
- *Isolation* refers to the fact that during concurrent transactions, each transaction is isolated from other transactions.
- *Durability* means that all data after a transaction is completed will be saved by the database and available even if in the event of a database failure and restart.

The ACID properties are described in a standard ISO/IEC 10026-1:1998 (revised in 2003) focused on Distributed Transaction Processing.

4.8.2 Non-relational Database

Representation

Non-relational databases are databases where the data storage and retrieval are not organized in tabular relations. They are often referred as NoSQL databases in contrast to the relational databases. These databases were mostly developed in the early 2000s by large IT companies processing online content, such as Google, Amazon, Yahoo, LinkedIn, and Facebook. The motivation arose from a limited scalability of relational databases managing very large volumes of online content. The scalability limitations came from satisfying the ACID properties that assured consistency of the data in the case of simultaneous transactions but limited the database installation to a single server and did not allow the use of distributed computing. NoSQL databases do not strictly follow the ACID properties which allow them to scale well with big data and utilize distributed computational environments, such as computer clusters and clouds. Among the well-known NoSQL databases, management systems are MongoDB, Redis, Apache CouchDB, and Cassandra.

NoSQL databases followed many highly heterogeneous models for organizing data. We describe next four of the main models found in NoSQL databases:

1. Key-value stores
2. Graph stores
3. Column stores
4. Document stores

Key-value stores: Consist of an associative array where each key is paired to a value as shown in Fig. 4.13. As an analogy, one could think about the data organization model as a folder with each file name being the key and the content of the file being the value. These databases scale well but are limited when complex queries need to be performed on the database content.

Graph stores: Organize data as nodes and edges (or connections) as shown in Fig. 4.14. It is assumed that data can be modeled and represented as graphs. This data organization model performs well on interconnected data. However, the scalability may be limited because navigating through a long path in a graph composed

Fig. 4.13 Key-value store   database

key	value
K1	{A, 36, true}
K2	{07, 63}
K3	{hello, 0}

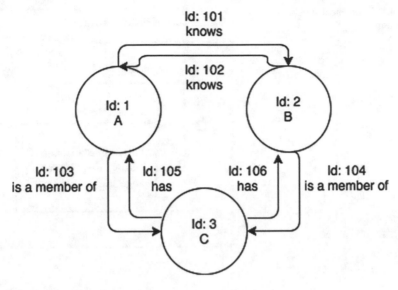

Fig. 4.14 Graph store database

First Name	Last Name	Age
John	Doe	36
Jane	Smith	44
Robert	Williams	63

> **Stored as:**
> John, Jane, Robert
> Doe, Smith, Williams
> 36, 44, 63

Fig. 4.15 Column store database

of distributed nodes could slow down the query responses. In the example below, vertices are entities (people A and B, organization C), and edges are relationships between these entities (A and B know each other and are members of C).

Column stores: Organize data in columns where each column represents a specific property of an entity (see Fig. 4.15). Generally, columns can be serialized on the same disk, making searches of any property fast.

Document stores: Store data in a semi-structured form, called documents (see a collection of documents in Fig. 4.16). Documents are usually encoded in a standardized format, such as eXtensible Markup Language (XML), JavaScript Object Notation (JSON), Binary JSON (BSON) or even Adobe Portable Document Format (PDF). Documents are not required to follow a specific schema, and therefore even entities of the same type might be represented differently.

Fig. 4.16 Document store
database

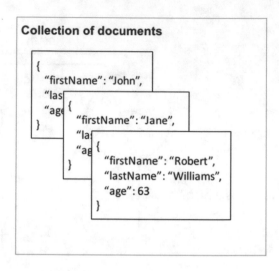

MongoDB database management system

One of the most widely used NoSQL DBMSs is MongoDB [35]. It is free, open-source, cross-platform, and document-oriented DBMS developed by MongoDB Inc. Documents are stored in a JSON-like format. They contain fields of a variety of types, such as string, numbers, arrays, binary data, or embedded documents. Documents are organized in collections, and the document model can vary from one document to another. Database drivers, software acting as a layer between the DBMS and software applications, are available for most programming languages, avoiding the need to create an ODM (Object Document Mapper) layer to map MongoDB documents into objects. The MongoDB management system also provides the ability to use indices for improving queries performances and security features with database authentication. MongoDB as a NoSQL database offers a flexible data organization model that allows developers to modify their data structure easily and scales with data volume better than relational databases.

4.8.3 Java Spring Framework for Web Application Development

Once information is stored in a database on a web server, a developer needs to write the code that exchanges information between the database, computational nodes, file system, and web clients. In Sect. 4.4, we introduce the communication protocols for these purposes and mentioned Java Spring Framework as the server application for handling the information exchange. Here, we provide more background on the role of Java Spring Framework as one of the web application frameworks.

Software engineering with web application frameworks
When building software and web applications, a good practice for development and maintainability is to assure that the code is properly structured and follows software engineering standards. To achieve that, web software developers can use application frameworks that provide generic functionality and facilitate development of well-structured software applications.

Specifically, web application frameworks provide interfaces for common operations during the development. The common operations include exchanging information with:

(a) Databases (data access layer)
(b) Applications (API)
(c) User interfaces (GUI creation)
(d) Software functionality tests
(e) A variety of integrated utilities

At the component level, the web applications follow object-oriented programming principles (class inheritance, interface implementation, abstraction of data and behavior, encapsulation of data and class implementation, polymorphism, and virtual methods[16]), and their description is outside of the scope of this book. At the web system level, web applications follow a client-server architecture (also called a software design pattern). Some examples of client-server architectures include the Model-View-Controller (MVC), Model-View-View-Model (MVVM), or three-tier architecture. The MVC architecture was described in Sect. 4.5.1. Some well-known open-source application and web frameworks are Django (Python framework), Spring (Java framework), Ruby on Rails (Ruby framework), Symphony (PHP framework), Qt (cross-platform framework), and AngularJS and BackboneJS (JavaScript frameworks). We will describe Spring Framework as one of the most used frameworks for building Java enterprise applications.

Overview of Java Spring Framework
The Spring Framework [36] provides a programming and configuration model for building Java applications. It also provides core functionality needed in enterprise web applications so that a developer can focus on the logic of the application code more than on the interfaces to other web application components. Figure 4.17 shows several groups of about 20 modules in Spring Framework.[17] The key characteristics of Spring Framework and its modules are the interfaces following the categories listed above. As shown Fig. 4.17, Spring Framework provides interfaces to databases (data access/integration modules), generation of web REST APIs and development of servlets for web applications (web modules), configuration of application components using dependency injection (core modules), and support for testing applications (test modules).

---

[16] http://www.introprogramming.info/english-intro-csharp-book/read-online/chapter-20-object-oriented-programming-principles/
[17] http://docs.spring.io/spring-framework/docs/current/spring-framework-reference/html/overview.html#overview-modules

Fig. 4.17 Overview of the
Spring Framework

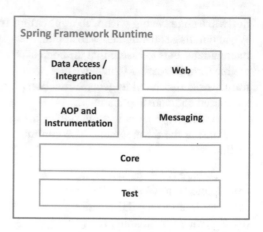

The collection of modules labeled "AOP and instrumentation" consists of methods following aspect-oriented programming (AOP) and those for interacting with application web servers, such as Apache Tomcat. AOP is a programming paradigm that increases modularity of software by cleanly decoupling code. The modules labeled "Messaging" make methods available for converting plain old Java objects (POJO) into messages and sending them to applications (message-driven architecture). The content can be Java messages sent to applications to trigger actions, for example, HTTP requests, or even transferring files.

The WIPP's design leverages two main Spring Framework modules from the data access/integration group and one Spring Framework project that are detailed below:

1. Spring Data MongoDB module
2. Spring Data REST module
3. Spring Boot project

Spring Data MongoDB

Spring Data MongoDB is used for object-document support and for integration of MongoDB with Spring object repositories. This Spring Framework module provides functions for interacting with MongoDB collections and for mapping the documents from a database into plain old Java objects (POJOs) used in computational and data management applications. All Spring Data modules use the concept of Spring Repositories for interacting with data stored in a database. Spring repositories offer interfaces to the database CRUD operations, search queries, and can map data stored in databases into Java objects or vice versa. It is the Spring repository in WIPP that manages all MongoDB collections and allows manipulation of the database entities.

Spring Data REST

Spring Data Representational State Transfer (REST) is used to export Spring Data Repositories as hypermedia-driven RESTful resources. This module provides an easy way for developers to create REST web services. REST web services expose the data in Spring Repositories as hypermedia-driven HTTP resources which are accessible by

URLs. The navigation is achieved via hyperlinks exposed with the resources pointing to multiple HTTP resources. Spring Data REST also embeds useful tools for REST API. For instance, the Spring Framework module supports (a) pagination of tabular results with navigational links to previous and next pages and (b) query-based search for HTTP resources in the Spring repositories. All HTTP resources are exposed as JSON documents following the HAL standard (Hypertext Application Language, for representing resources and their relations with hyperlinks).

Spring Boot
Spring Boot [37] is a Spring Framework project to facilitate an easy creation of Spring-based Java applications. It dynamically adds and pre-configures Spring Framework modules by scanning the developed application. For example, Spring Boot embeds the Apache Tomcat web server for deploying a web application. By using Spring Boot, a developer can enable external configurations through property files. These automated configurations of Spring modules can make the developer more productive by allowing him to focus on the code that is specific to his application domain.

4.9   Meeting Computational Requirements on a Web Server

Another key components of WIPP is the computational scalability of image processing operations in distributing computing environments. To meet the computational requirements, WIPP is leveraging one of the scientific workflow management systems called Pegasus. The topic of designing scientific workflow management system is its own research area and readers are referred to research papers, such as [38–40]. In this section, we focus on Pegasus, its HTCondor workload management system, and the XML file representation for encoding computational workflows.

Scientific workflow management systems
Workflow management systems (WMS) are software applications that allow the execution and monitoring of a variety of computational tasks. Workflows of tasks are generally represented as graphs or pipelines. In a scientific application, workflows perform tasks that manipulate data or compute new data. Scientific WMS have been developed to lower the barriers for scientists to utilize rapidly changing hardware. Scientists can construct a workflow of computational tasks from data and tools, execute a workflow, monitor its execution, and retrieve the results along with provenance information. Widely used WMS in multiple scientific communities include Apache Taverna, Kepler, Galaxy, KNIME, and Pegasus.

Pegasus workflow management system
Pegasus WMS is a workflow management system used in WIPP. It integrates several middleware libraries including HTCondor so that Pegasus can offer many desirable features of WMS. The interface between Pegasus WMS on a server side and user's inputs on a client side is facilitated via an XML file representation. We provide additional descriptions of Pegasus, HTCondor, and XML file representation next.

Fig. 4.18 Example of a
workflow DAG. Input data
is separated into two sets
and processed in parallel,
and results are merged to
deliver the output data

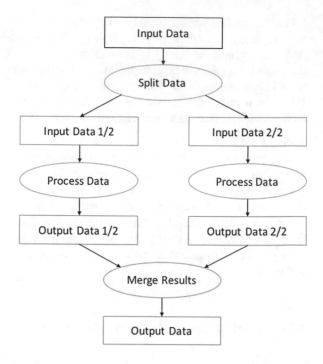

4.9.1 Pegasus Workflow Management System

Pegasus is specifically designed for creation, execution, reuse, and repurposing of
scientific workflows in a variety of execution environments. Workflows are represented as Directed Acyclic Graphs (DAGs) of tasks where each node in the graph is
a task and each edge is a dependency between tasks. Example of a workflow represented by a DAG is shown in Fig. 4.18. Abstract workflows represented as DAGs
are stored in XML format in the files with .DAX suffix (DAX stands for Description
of Abstract workflow in XML).

Abstract workflows stored as DAX files
DAX workflows describe a set of tasks (also called jobs) and their dependencies.
Each job is mapped to an executable and an execution site. It can be associated with
a set of input arguments, as well as input and output files. Three catalogs are used in
the Pegasus WMS and are referenced in a DAX file:

(a) Sites catalog: It describes the execution sites that are available for the execution
   of the jobs.
(b) Transformations catalog: It describes the available executables (software applications or programs).
(c) Replica catalog: It describes the mapping between logical file names used in the
   DAX to physical file locations.

User interface
For a user, Pegasus can be viewed as an interface between a domain scientist and the
computational environments. High-level workflow descriptions are mapped to a

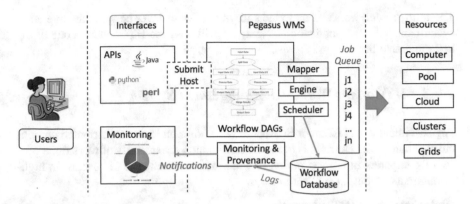

Fig. 4.19 Overview of Pegasus WMS

sequence of computational and data management tasks. The tasks are then adapted
to the specific computational software and hardware environment to maximize efficiency of the workflow execution. Computation execution environments can vary
from a simple desktop computer to a computer grid, cluster, or cloud (e.g., National
Science Foundation and Department of Energy funded Open Science Grid[18] and the
Extreme Science and Engineering Discovery Environment[19] (XSEDE)).

Developer interface
For a developer, Pegasus provides several APIs for creating workflows in different
programming languages (Python, Java, and Perl) and a REST API for monitoring
workflow executions. While utilizing Pegasus, one must understand the underlying
three parts of the Pegasus WMS as illustrated in Fig. 4.19.

1. Workflow Mapper converts an abstract workflow to an executable form while
   performing an optimization of the workflow, adding data staging and registration
   tasks, and selecting resources.
2. Workflow Engine executes the concrete workflow and submits jobs to the job
   scheduler.
3. Job Scheduler schedules the submitted jobs on the selected resources.

As a workflow management system used in WIPP, Pegasus adds value to image
processing by including:

(a) Portability and reuse of workflows with abstract DAXs
(b) Workflow performance optimization
(c) Computational scalability on distributed hardware
(d) Gathering provenance information
(e) Data management
(f) Reliability and error recovery of computations.

Many of these additional Pegasus execution characteristics come from integrated
middleware libraries, such as HTCondor, Globus, or cloud API to the Amazon EC2.

---

[18] https://www.opensciencegrid.org/

[19] https://www.xsede.org/

The term "middleware" refers to the software services layer between the operating system and software applications. Next, we will describe HTCondor as one of the key integrated libraries in Pegasus.

4.9.2   HTCondor Workload Management System

Pegasus integrates HTCondor to execute workflows on local or distributed computational hardware. The goal is to manage efficiently the quantity of work done by a set of computers in a given period of time. This efficiency is of concern in high-throughput computing applications.

High-throughput (HT) computing

HTCondor is an open-source distributed computing software for high-throughput computing applied to large collections of data over distributed computational resources.[20] It is designed for workload management of compute-intensive jobs, job scheduling and queueing, as well as monitoring and management of computational resources. HTCondor can utilize a single computer or a pool of hardware resources. In this case, the Condor pool is the place where each machine advertises its available resources, such as CPU type and speed, size of RAM memory, size of virtual memory, and currently executing workload. This information is used by the HTCondor scheduler to send jobs to computational resources that have the needed specifications and are available to execute the jobs.

HTCondor currently supports several execution environments, called "universes":

- Standard universe: As the default environment, it provides reliability and ease of migration. However, it is constrained by the inability to use multiprocess jobs.
- Vanilla universe: It has fewer restrictions but also fewer features than the standard universe.
- Grid universe: It allows users to submit their jobs on computational grids.
- VM universe: It supports launching virtual machine disk images instead of programs.
- Docker universe: It can launch Docker containers instead of programs.

When Pegasus WMS is running on top of HTCondor, the Vanilla universe is used due to the few restrictions on this universe.

4.9.3   XML File Representation for Encoding Computational Jobs

We use the XML (eXtensible Markup Language) format as the common representation for the input parameters of jobs in WIPP. For each job, input parameters are configured by the user using the web interface. Some job inputs are generated by the WIPP system on the server, for example, a unique output folder path. All parameters

---

[20] http://research.cs.wisc.edu/htcondor/

```
<StitchingWork>
    <id>xxxxxxxx-xxxx-xxxx-xxxx-xxxxxxxxxxxx</id>
    <tilesFolder>/data/WIPP/image-collections/yyyyyyyy-yyyy-yyyy-yyyy-yyyyyyyyyyyy</tilesFolder>
    <outputFolder>/data/WIPP/tmp/jobs/xxxxxxxx-xxxx-xxxx-xxxx-xxxxxxxxxxxx</outputFolder>
    <algorithm>MIST</algorithm>
    <options>
        <filenamePatternType>ROW_COLUMN</filenamePatternType>
        <filenamePattern>Well5_Pos-c{cc}-r{rr}_t{ttt}.ome.tif</filenamePattern>
        <startingPoint>TOP_LEFT</startingPoint>
        <gridWidth>10</gridWidth>
        <gridHeight>10</gridHeight>
        <horizontalOverlap>10</horizontalOverlap>
        <verticalOverlap>10</verticalOverlap>
        <stageRepeatability>5</stageRepeatability>
    </options>
</StitchingWork>
```

Fig. 4.20 Example of an XML input file for a stitching vector estimation job

are stored in the database and serialized to an XML file. The XML file is then sent by Pegasus WMS to a specific software implementation on the server that can parse the inputs needed for the execution.

An example of XML input parameters for a stitching job is shown in Fig. 4.20. In this example, a user chose to compute a stitching vector using the MIST algorithm [41] and provided specific parameter values to achieve optimal algorithmic performance. The algorithm is applied to image tiles in a specified collection. To identify spatially adjacent pairs of image tiles, a user provided a file name pattern for extracting the time frame and row and column position from each image file name. The file name pattern is defined with a regular expression understood by the MIST algorithm. From the collection of images, the WIPP system generates the parameter "tilesFolder" that encodes a path of this input collection following its unique identifier in the data storage system. The web interface assists a user with job configurations. It has embedded default values for the parameters and runs JavaScript code in the browser to pre-populate some of the fields automatically after scanning the image file names in the input collection.

4.10 Delivering Traceable Computations

As WIPP is configured to manage all metadata about computational jobs, it can also serve as a repository of provenance information. To benefit from the provenance information and deliver traceable computations, additional components must be designed. In this section, we describe how traceability of statistical summaries is achieved in WIPP.

Imaging traceability

We are interested in traceability of quantitative and qualitative image object measurements that lead to discoveries or decision-making. From a measurement perspective, metrological traceability requires the establishment of an unbroken chain of calibrations to specified references.[21] This traceability definition is important for preparing and imaging specimens so that the measured images can be related to a traceable standard or any other reference with known uncertainty.

Computational traceability
In computer science, traceability is defined as the "ability to relate artifacts created during the development of a software system to describe the system from different perspectives" [42]. In the context of WIPP and digital image analyses, the traceability is understood as a chain of image transformations and computations that have been applied to acquired images through the entire software processing workflow. Traceable image files can be audited for correctness and completeness at any time point using provenance information. The provenance information contains metadata about data, software, and hardware. The access to provenance information is conveniently offered to end users via the WIPP web interface and step-by-step browsing of an entire sequence of processing tasks.

Next, we will describe the WIPP components for delivering traceable computations and web statistical modeling module in WIPP that assists in publishing traceable statistical summaries. Finally, we will discuss the challenges of transitioning from traceable to reproducible computations.

4.10.1 Components for Delivering Traceable Computations

Traceable computations are delivered in WIPP by hyperlinking each image collection and each computational job with its inputs and outputs. To achieve traceability, four basic components of WIPP play a role in gathering, storing, querying, retrieving, and browsing provenance information:

- Database stores provenance information about data artifacts and jobs configurations.
- REST API is used for querying provenance information about all objects stored in the WIPP database. The objects are exposed through REST endpoints as web resources (i.e., URLs).
- Pegasus WMS gathers computational provenance information about job executions, and updates the WIPP system database via the REST API using its notification system during each job execution.
- Web interface is used for retrieving and interacting with provenance information via dynamic web pages. A user can browse provenance information retrieved from the WIPP database via the REST API.

Example
If an image collection was generated as a job output, then the job description is hyperlinked from the detailed view of the generated image collection. Similarly, if an image collection is an input to a job, then this collection will be hyperlinked from

---

[21] https://www.nist.gov/traceability/nist-policy-metrological-traceability

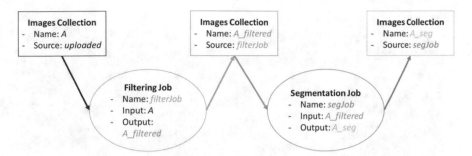

Fig. 4.21 Example of an image processing pipeline with filtering and segmentation jobs. Each computational step is linked with its input and output

the detailed view of the computational job. A user can follow hyperlinks from the viewed image collection to recover all computational jobs that have been applied to an uploaded image collection to generate the viewed collection. Figure 4.21 illustrates a sequence of filtering and segmentation jobs that generates two new image collections (A_filtered and A_seg) from the uploaded collection A. By starting at any place of this sequence, a user can trace a chain of computations to the uploaded collection A.

4.10.2   Traceable Computations for Publications

Traceable histograms
Traceability is important for scientists when publishing and sharing their findings. To facilitate traceability of published data, the WIPP system contains the web statistical modeling (WSM) module. This WSM module is accessible after building an image pyramid, segmenting the image, and extracting features for each image segment from the feature extraction module. Given the pyramid and feature values derived from the same input image collection, a user can interactively explore histograms of objects based on each feature. The histograms are interactively filtered by object spatial locations and by feature values. When filtering is completed and a user would like to publish the final histogram, then the histogram can be saved as an HTML document. Figure 4.22 shows the user interface for publishing a histogram (red box).

Histogram composition
The saved zip file contains a histogram HTML file that is rendered in Fig. 4.23. It contains a thumbnail of each object contributing to a histogram bin. Each thumbnail is hyperlinked to a Deep Zoom pyramid location as illustrated on the right side of Fig. 4.23. In addition, the saved zip file contains two additional files with the content shown in Fig. 4.24. One file provides all information about Deep Zoom coordinates and hyperlinks of thumbnails. Another file describes the histogram settings. By sharing the image pyramid on the Internet and publishing a scientific paper with the

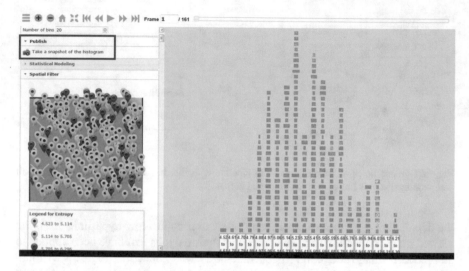

Fig. 4.22 The publish user interface in web statistical modeling tool

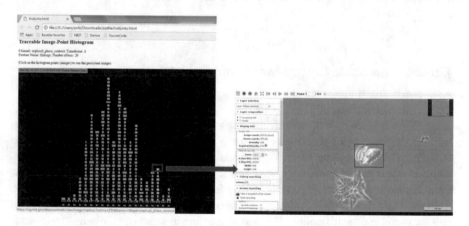

Fig. 4.23 Left – rendering of the saved HTML file from the web statistical modeling tool. Right – rendering of the web page accessed by clicking on one of the thumbnails in the histogram shown on the left side

histogram in an electronic journal, the saved zip file enables every reviewer and every reader to verify every point of the histogram result. Furthermore, if the histogram inputs (image pyramid and object features) are shared with their provenance information, then anyone viewing a scientific publication with the inserted histogram can trace all computations back to the uploaded image collection.

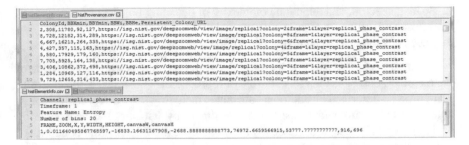

Fig. 4.24 Two additional files saved from the web statistical modeling tool that contain all pyramid coordinates and hyperlinks of thumbnails (top) and all information about the histogram parameters

4.10.3 From Traceable to Reproducible Computations

The traceability of computations is necessary but not sufficient condition for reproducibility of computations. Traceability is necessary because one must know input datasets, tasks configurations, software versions, and computational platforms to properly reproduce any computational result. Even if all the inputs and configurations are known, reproducibility can be difficult to achieve when going from one execution environment to another, particularly when dealing with different hardware platforms and operating systems. For example, 32-bit and 64-bit systems can yield very different results due to their different ways of handling floating point numbers. Furthermore, installing the same version of a software package on two systems may return slightly different results. The reason lies in many software packages relying on other external libraries with multiple versions. Thus, installations of the same software package on two systems might have different versions.

Using a centrally deployed software application such as WIPP might help with reproducibility of computations if WIPP is properly managed. First, WIPP provides traceable computations. Algorithm versions are known and fixed between new releases. Users can always check which software and its versions are used for their computations. Software updates might be a drawback for a user because new versions of an algorithm would not be available immediately. Users would have to wait for the next release of the web system before being able to use it.

4.11 Summary

As introduced in Chap. 1, big image data experiments require image measurements that address:

1. A wide range of physical and digital scales (nm to cm physical scale, TB- to petabyte (PB)-sized digital datasets)

2. Spatial, spectral, and temporal complexity of detecting objects of interests (cell mitosis, migration, apoptosis, differentiation)
3. Speed of time-critical computations (image analyses faster than cell changes)

After addressing scale, complexity, and speed, one obtains a numerical value of the measurement. Nonetheless, in the era of reproducible data science, the measurement must be described by additional attributes including trusted, traceable, repeatable, searchable, immutable, persistent, verifiable by humans, and accessible by multiple parties. These measurement attributes can be provided by functionalities of a web image processing pipeline.

In this chapter, we mapped functionality of a web image processing pipeline to information technologies in a client-server system. The technologies were classified into those running on a server side or a client side, and those responsible for client-server communication. Given a client-server system, we addressed visualization of large images, as well as storage and management of large image, to enable images to be verifiable by humans and accessible by multiple geographically distributed parties. Finally, we described technologies that deliver three more measurement attributes, such as (1) scalability, (2) traceability, and (3) searchability via workflow management systems, databases, and web technologies.

The chapter provides basic concepts of multiple information technologies without providing examples to develop programming skills. The reason lies in a plethora of online materials and online "playgrounds" for gaining hands-on experience. After completing this chapter, readers should be able to make informed decisions about a source of problems in WIPP if it occurs. The chapter also offers information for those who would like to modify underlying WIPP technologies to meet additional big image data requirements.

References

1. MongoDB: [Online]. Available: https://docs.mongodb.com/. (2016) Accessed 22 July 2016
2. Allan, C., et al.: OMERO: flexible, model-driven data management for experimental biology. Nat. Methods. **9**(3), 245–253 (2012)
3. Spring Framework: [Online]. Available: http://projects.spring.io/spring-framework/ (2015)
4. Rasband W.: ImageJ & Fiji & ImageJA & ImageJ2, Computer Program. [Online]. Available: http://rsbweb.nih.gov/ij/ (2013)
5. Oracle: Java Advanced Imaging. On-line Documentation. [Online]. Available: http://www.oracle.com/technetwork/java/javase/tech/jai-142803.html (2017). Accessed 14 Sep 2017
6. Blattner, T., Keyrouz, W., Chalfoun, J., Stivalet, B., Brady, M., Zhou, S.: A Hybrid CPU-GPU System for Stitching Large Scale Optical Microscopy Images. in 2014 43rd International Conference on Parallel Processing, pp. 1–9. (2014)
7. Pegasus workflow management system: [Online]. Available: https://pegasus.isi.edu/ (2017). Accessed 08 Jan 2017
8. HT Condor: workload management system for compute-intensive jobs. Unversity of Wisconsin-Madison. [Online]. Available: https://research.cs.wisc.edu/htcondor/ (2017). Accessed 14 Sep 2017
9. Google: AngularJS. [Online]. Available: https://angularjs.org/ (2017). Accessed 14 Sep 2017

10. Open Seadragon: Open Seadragon project. [Online]. Available: http://openseadragon.github. io/ (2017). Accessed 24 Feb 2017
11. CamanJS: (ca)nvas (man)ipulation in Javascript. [Online]. Available: http://camanjs.com/ (2017). Accessed 14 Sep 2017
12. WebDeepZoomToolkit: [Online]. Available: https://github.com/usnistgov/WebDeepZoom Toolkit (2017). Accessed 14 Sep 2017
13. Berson, A.: Client-Server Architecture, 2nd edn. McGraw-Hill, New York (1996)
14. Smith, P.N., Guengerich, S.L.: Client/Server Computing, 2nd sub-Ed. Sams Publishing (1994)
15. Sommerville, I.: Software Engineering, 10th edn. ©2006: Addison-Wesley Longman Publishing Co., Inc., Boston, MA (2015)
16. Jacobs, I., Walsh, N.: URI/Resource Relationships. Architecture of the World Wide Web, Volume One (2004)
17. Totty, B., Gourley, D., Sayer, M., Aggarwal, A., Reddy, S.: HTTP: The Definitive Guide. O'Reilly Media, California (2009)
18. Kozierok, C.M.: The TCP/IP Guide: A Comprehensive, Illustrated Internet Protocols Reference, 1st ed. No Starch Press (2005)
19. Snir, M., Otto, S., Huss-Lederman, S., Walker, D., Dongarra, J.: MPI: the complete reference. In: Kowalick, J. (ed.) Scientific and Engineering Computation Computation, p. 350. MIT Press, Cambridge, MA (1996)
20. Callaghan, B.: NFS Illustrated, 1st ed. Addison-Wesley Professional (2000)
21. Felke-Morris, T.: Basics of Web Design: HTML5 & CSS3, 3rd edn. Pearson (2015)
22. Zakas, N.C.: Professional JavaScript for Web Developers, 3rd edn. Wrox (2012)
23. Keith, J.: DOM Scripting: Web Design with JavaScript and the Document Object Model, 2nd ed. Apress (2010)
24. Borini, S.: Understanding Model-View-Controller. GitBook.com (2017)
25. Deep Zoom Silverlight. Microsoft Developer Network (MSDN). [Online]. Available: http:// msdn.microsoft.com/en-us/library/cc645050(v=vs.95).aspx (2017). Accessed 14 Sep 2017
26. Saalfeld, S., Cardona, A., Hartenstein, V., Tomančák, P.: The Collaborative Annotation Toolkit for Massive Amounts of Image Data (CATMAID). Max Planck Institute of Molecular Cell Biology and Genetics. [Online]. Available: http://catmaid.org/ (2017). Accessed 14 Sep 2017
27. Outercurve Foundation: Cosmic chronology ChronoZoom. web page. [Online]. Available: http://www.chronozoomproject.org/BehindTheScenes.htm (2017). Accessed 14 Sep 2017
28. Vandecreme, A., et al.: From image tiles to web-based interactive measurements in one stop. Micros. Today. **25**(1), 18–25 (2017)
29. Bajcsy, P., et al.: Enabling interactive measurements from large coverage microscopy. IEEE Comput. **49**(7), 70–79 (2016)
30. Kooper, R., Bajcsy, P., Hernández, N.M.: Stitching giga pixel images using parallel computing. In: IS&T/SPIE Electronic Imaging, pp. 7872–17 (2011)
31. Kooper, R., Bajcsy, P.: Computational scalability of large size image dissemination. In: IS&T/ SPIE Electronic Imaging, San Francisco, CA, pp. 7872–23 (2011)
32. Bajcsy, P., Vandecreme, A., Amelot, J., Nguyen, P., Chalfoun, J., Brady, M.: Terabyte-sized image computations on hadoop cluster platforms. In: IEEE International Conference on Big Data, Santa Clara, CA (2013)
33. Prigmore, M.: Introduction to Databases with Web Applications. Prentice Hall, Harlow (2007)
34. Williams, H.E., Lane, D.: Web Database Applications with PHP & MySQL. O'Reilly Media (2004)
35. Chodorow, K.: MongoDB: The Definitive Guide, 2nd edn. O'Reilly Media (2013)
36. Walls, C.: Spring in Action: Covers Spring 4, 4th edn. Manning Publications (2014)
37. Walls, C.: Spring Boot in Action, 1st edition. Manning Publications (2016)
38. Ludäscher, B., et al.: Scientific workflow management and the kepler system ∗. Electr. Eng. **78296**(10), 1–19 (2005)

39. Li, X., Song, J., Huang, B.: A scientific workflow management system architecture and its scheduling based on cloud service platform for manufacturing big data analytics. Int. J. Adv. Manuf. Technol. **84**(1–4), 119–131 (2016)
40. Rodriguez, M.A., Buyya, R.: Scientific workflow management system for clouds. In: Mistrik, I., Bahsoon, R., Ali, N., Heise, M., Maxim, B. (eds.) Software Architecture for Big Data and the Cloud, 1st edn, pp. 367–387. Elsevier/Morgan Kaufmann publisher, Cambridge, MA (2017)
41. Chalfoun, J., Majurski, M., Blattner, T., Keyrouz, W., Bajcsy, P., Brady, M.: MIST accurate and scalable microscopy image stitching method with stage modeling and error minimization. Nat. Sci. Reports. **7**, 1–10 (2017)
42. Spanoudakis, G., Zisman, A.: Software traceability: a roadmap. In: Handbook of Software Engineering and Knowledge Engineering, pp. 395–428. World Scientific Publishing, Singapore (2004)

Chapter 5
Image Processing Algorithms

5.1 Inputs and Outputs of Algorithms

Let us focus on image processing algorithms that convert a collection of raw microscopy images (inputs) into object measurements (outputs). A collection of raw microscopy images is assumed to represent multiple spatially overlapping fields of view (FOVs), acquired over several spectral wavelengths and/or time points. The raw FOVs can be assembled into a large 2D or 3D image with one or multiple spectral values over time (2D/3D space +1D time + nD spectrum).

For example, co-registered phase contrast and fluorescent FOVs of cell colonies through time can form 2D gigapixel-frame terabyte-sized videos with two spectral values per [x, y, time] coordinate. There are three such video examples found online[1] with phase contrast and fluorescent spectral values at each [x, y, time]. The physical dimensions of the input collection are 2D in space, 1D in time, and 2D in spectra. In this example, algorithms convert 359 568 raw input FOVs (image files) into cell colony measurements where cell colonies grow and merge over time. The algorithms are designed to ingest images and output processed images (e.g., flat-field corrected and segmented images), metadata about input images (e.g., stitching vector), or tabular data containing measurements derived from images (e.g., cell colony size).

In this chapter, inputs to image processing algorithms are image collections, and the outputs are either processed image collections or image measurements. The image collections contain images with gigapixels (10^9 pixels), and their storage reaches a terabyte size (10^{12} bytes). Our next step is to introduce the design of image processing algorithms.

---

[1] https://isg.nist.gov/deepzoomweb/data/stemcellpluripotency

© Springer International Publishing AG 2018
P. Bajcsy et al., *Web Microanalysis of Big Image Data*,
https://doi.org/10.1007/978-3-319-63360-2_5

5.2   Image Processing

Before we focus on several algorithms included in WIPP, we briefly introduce image processing via available textbooks and image processing software. Image processing theory has been documented in many textbooks, and they are the best starting point for a novice. To gain practical understanding of the theory, one can install several image processing tools and apply them to acquired microscopy images. We support this educational process by providing pointers to image processing textbooks, classifying image processing software based on its usage, and illustrating the use of image processing operations throughout the rest of this section.

5.2.1   Textbooks About Image Processing

For a reader who is new to image processing, we briefly refer to four example textbooks that introduce basic concepts about:

1. Image processing [1]
2. Image processing [2]
3. Microscope image processing [3]
4. Medical image processing [4]

Basic concepts about image processing
From the image processing textbook [1], the reader can learn about digital image fundamentals; intensity transformations and spatial filtering; filtering in frequency domain; restoration and reconstruction; color image processing; wavelets and multi-resolution processing; image compression; morphological processing; image segmentation, representation, and description; and object recognition. The handbook [2] includes additional aspects of printing, defect correction, feature measurement, and 3D acquisition and processing. To derive quantitative measurements from images, these fundamentals must be enhanced by our knowledge about (a) the microscope instruments generating digital images, (b) the sample under the microscope, and (c) the image analysis task in the context of a specific biomedical application.

Microscope image processing
The book on microscope image processing [3] expands on the subject of image processing and includes the fundamentals of microscopy, image digitization, image display, geometric transformations, image enhancement, wavelet image processing, morphological image processing, image segmentation, object measurements, object classification, fluorescence imaging, multispectral imaging, 3D imaging, time-lapse imaging, autofocusing, structured illumination imaging, and image data management. Most of the topics revolve around understanding microscope-specific image generation which informs the process of deriving quantitative measurements. These topics include quantitative definition of image, pixel, and color in physical units,

introducing point spread function, Abbe's diffraction limit, and Nyquist-Shannon sampling theory. Finally, quantitative data derived from image measurement is often validated and verified using human inspection. To understand human inspection challenges, image display, human perception, and reproducibility of measurements are also addressed.

Medical image processing

A reader interested in microscope image processing can learn much from textbooks in medical image processing. In the medical image processing handbook [4], the topics cover image enhancement, segmentation, quantification, registration, visualization, compression, storage, and communication. While these topics seem similar to digital image fundamentals, they are presented in the medical context of diagnosing disease in human organs imaged by magnetic resonance, positron emission tomography, computed tomography, X-ray, and ultrasound imaging modalities. The general image processing principles are linked to the medical applications in this handbook.

5.2.2 Usage-Based Classification of Image Processing Implementations

Classification based on software usage

After learning about image processing from textbooks, one can apply existing image processing implementations to solve specific problems. The spectrum of existing software packages, libraries, and scripting and scientific workflow environments for building image processing solutions is surveyed in [5, 6]. Table 5.1 summarizes how each type of image processing implementation would be used. One should be aware that the types of implementations in Table 5.1 have a specific meaning in the context of their use.[2] We describe the types in more details next.

Table 5.1 The usage of software implementations for image processing operations

Type of image processing implementations	Usage
Software package	Apply existing operations to a set of predefined tasks without changes (no programming)
Software library	Create a solution for a set of new tasks by calling existing operations programmatically and adding/modifying the software implementations
Scripting environment	Script existing operations using a scripting language to create a solution for a set of new tasks
Workflow environment	Integrate and script existing operations with minimum programming, and leverage predefined support for efficient executions on a set of computational hardware

---

[2] http://www.farsight-toolkit.org/wiki/Main_Page

Software package and libraries

A software package is a stand-alone, tightly integrated system that is used for a predefined set of tasks. *A software library* is a collection of implementations with a well-defined application programming interface (API). A software library has a higher potential for being reused than a software package. By exposing a well-defined API in a software package, the package can become a library or a toolkit for building customized solutions. Due to the possible unintended uses of software functionality and their legal ramifications, software licenses for toolkits typically contain special clauses about their reuse. Examples of software packages are Visiopharm[3] (cancer diagnostics), Indica Lab[4] (pathology tissue classification tasks), or lmaris[5] (tasks related to data management, visualization, analysis, segmentation, and interpretation). Examples of open-source libraries are the Visualization ToolKit (VTK)[6] (processing and visualization), the National Library of Medicine Insight Segmentation and Registration Toolkit (ITK)[7] (segmentation and registration), and Open Source Computer Vision Library (OpenCV)[8] (computer vision-related operations).

Scripting environments

A scripting environment is a collection of standardized APIs for accessing software library implementations within a scripting language. Scripting environments facilitate the automation of processing. The scripting language may be different from the library's core programming language to make the library accessible to a broader user community. Examples of specialized scripting environments are R (statistical analyses), LabVIEW (analyses linked to instrumentation acquisition), or MATLAB (mathematical analysis). Examples of general-purpose scripting languages are JavaScript, Python, and Perl.

Scientific workflows

Scientific workflow environments [7] have been developed to assist with:

(a) Software organization
(b) Interactions of software written in multiple programming languages
(c) Reusability of software
(d) Community software sharing
(e) Gathering provenance
(f) Security
(g) Scalability of execution
(h) Built-in fault tolerance

We are specifically referring to scientific workflow systems as a subclass of general workflow management systems used for the automation of business processes (e.g.,

---

[3] https://www.visiopharm.com/

[4] http://www.indicalab.com/

[5] http://www.bitplane.com/imaris/imaris

[6] http://www.vtk.org/

[7] http://www.itk.org/

[8] http://opencv.org/downloads.html

an online business process of purchasing a flight ticket followed by a car rental reservation). Workflow environments for image processing can lower the bar for scripting via a visual programming interface which allows for dragging and dropping an image processing operation into a workflow pane and connecting it with other operations using a toolbar and mouse clicks. Examples of open-source scientific workflows used for microscopy and computational biology analyses are KNIME,[9] ilastik,[10] and Galaxy[11].

Notes about software implementation types
Many implementations fall into more than one software implementation category. For example, ImageJ can be used as a software package, a library, or a scripting environment that supports scripting languages, such as ImageJ macro, JavaScript, BeanShell, Jython (Java implemented Python), JRuby (Ruby implemented in Java), Clojure, and Groovy. A software package can also be turned into a library, the same way as a scripting environment can be enhanced to become a scientific workflow. However, the "best" implementation should be determined in terms of (a) the productivity of end users when solving tasks and (b) the combination of the largest set of trusted image processing implementations within a working environment that offers the most user-friendly software. Finally, the term "community platform" has been used by many software projects when image processing libraries are bundled together with scientific workflow functionalities and with image databases. We have not included image databases in our usage-based classification because from a user perspective, the storage and representation of data are of concern only for developers of new image processing algorithms.

5.2.3   Classification of Open-Source Image Processing Software

Searching for implementations
After gaining enough knowledge about image processing and choosing the preferred mode of use, one begins searching for existing implementations. To assist in this search process, we consolidated the surveys in [5, 6] and presented their software classifications together with our usage-based categories. According to [6], bioimage analysis tools used for computer-based image analysis can be classified into those with command-line interface (CLI), graphical user interface (GUI), scripting language interface, and database interface. According to [5], the primary functions of such software are workflow, image analysis, machine learning, and image acquisition.

Classification of implementations
Table 5.2 presents a summary of all three types of implementations. We did not include commercial software because most of the implementations in the surveys

---

Table 5.2 Summary of existing open-source implementations of image processing

Software name	Usage-based classification	Primary function	Interface-based categories
Bio7	Scientific workflow	Workflow system	GUI tools: Generic platform
Bio-formats	Library	Image format conversion library	CLI
BioImageXD	Software package, library, scripting environment	Image analysis	GUI tools: Generic platform
BisQue (bio-image semantic query user environment)	Scientific workflow	Image database	Image databases
CellCognition and VIGRA	Scientific workflow, library	Image analysis	Specialized software
CellProfiler and analyst	Scientific workflow, library	Machine learning and data analysis	GUI tools: Workflow-based
CellOrganizer	Software package, library	Machine learning, modeling, and visualization	Specialized software
FarSight	Library, scripting environment	Visualization	GUI tools: Workflow-based
Icy and ImageJ, VTK, micro-manager	Scientific workflow, library	Image analysis	GUI tools: Generic platform
Ilastik	Scientific workflow	Machine learning	GUI tools: Workflow-based
Image processing in R	Scripting environment	Scripting environment	CLI
ImageJ/Fiji	Software package, library, scripting environment	Image analysis	GUI tools: Generic platform
ITK	Library	Bioimaging library	CLI
KNIME	Scientific workflow	Workflow system	CLI
Lever	Software package, library	Image analysis	Specialized software
NeuroStudio	Software package, library	Image analysis	Specialized software
OMERO	Software package, library, scripting environment	Image database	Image databases
OpenBIS	Scientific workflow	Workflow system	Image databases
OpenCV (open source computer vision library)	Library	Bioimaging library	CLI
PSLID (protein subcellular localization image database)	Software package, library	Machine learning	Image databases

(continued)

Table 5.2 (continued)

Software name	Usage-based classification	Primary function	Interface-based categories
Python and ski image	Scripting environment and library	Scripting environment	CLI
ScanImage	Software package, library	Image acquisition	Specialized software
TMARKER	Software package, library	Image analysis	Specialized software
Vaa3D	Software package, library	Visualization and image analysis	GUI tools: 4D + t data exploration and analysis
VTK	Library	Bioimaging library	CLI
WND-charm	Library	Machine learning	CLI
μManager	Scripting environment and library	Image acquisition	CLI

GUI stands for graphical user interface. *CLI* stands for command-line interface.

were open source. In addition, open-source software has available web information for determining their classification, while the commercial web pages do not. It can also be seen from Table 5.2 that there is no simple classification method for software given its many uses, various primary functions, and many access interfaces. The primary software functionality can provide guidance as to where it might be most usefully incorporated into WIPP or any other image processing system.

Challenges of integrating multiple implementations
Unfortunately, many desktop-based visualization and workflow construction systems must be rewritten to run in a client-server system. The interface-based categories can be helpful in assessing the complexity of software rewrite and integration. For example, software with command-line interfaces are easier to integrate than the software with GUI. The usage-based classification helps select the right software. The web image processing pipeline aims at offering access to a broad spectrum of software implementations (libraries), scripting mechanisms using web RESTful services, and scientific workflow systems via web interfaces. For instance, several image database solutions listed in Table 5.2 already take advantage of web systems (e.g., OMERO, BisQue) because they must address the storage and computational challenges of big data.

5.2.4 Loading Images Using OME Bio-Formats Library

After choosing the image processing implementations with the desired functionalities, the initial challenge lies in loading images. This challenge arises due to the many file formats used by microscope and camera vendors and the need for having the ability to read not only image data but also the acquisition parameters stored in the files. To address this challenge, the Open Microscopy Environment (OME[12]) was established in 2011 as a collaborative effort among academic and commercial entities with funding provided by the Wellcome Trust Strategic Award. OME produces open-source software for data management in biological light microscopy. As one of the software efforts, the Bio-Formats[13] library was developed to read and convert over 140 file formats to the OME-TIFF data standard.[14] We briefly present the OME-TIFF standard and the integration of the library for loading OME-TIFF image files.

Open Microscopy Environment TIFF image standard
The OME-TIFF image standard combines the pixel-level information stored in Adobe TIFF file format with the image-related information stored in OME-XML file format. A dataset in an OME-TIFF format has the following storage characteristics important for very large images:

- Images are stored within one multipage TIFF file or across multiple TIFF files. The ability to split large images across multiple files is advantageous during acquisition and data processing.
- The OME-TIFF file format supports the newer Big TIFF file extensions allowing the use of 64-bit byte offsets. It overcomes the file size limit in the original TIFF format with 32-bit byte offsets (limited to 4 GB file size).
- A complete OME-XML metadata block is embedded in each TIFF file's header. The OME-XML metadata block may contain any information represented as a < key, value > pair in a standard OME-XML file. Keeping the OME-XML metadata embedded in every file header introduces information redundancy and improves fault tolerance. An option was added to store partial OME-XML metadata blocks in each TIFF file's header with a reference to a master file containing the full OME-XML metadata. This allows a trade-off between the levels of redundancy and fault tolerance.
- OME-TIFF does not use OME-XML for encoding pixels as base64 chunks within the XML, and therefore OME-TIFF is preferred if there is at least one image in a dataset. The base64 encoding of images is used sometimes in URI for client-server communication. However, it is efficient only for very small images and not suitable for applications using OME-TIFF, such as high-content screening, time-lapse imaging, digital pathology, and other complex multidimensional image formats.

Note about base64 encoding The base64 encoding takes a stream of 8-bit pixel values, chunks them into 6-bit-long segments, and encodes them into 62 ASCII characters containing A–Z, a–z, and 0–9. The 63rd and 64th ASCII characters are defined depending on the specific variant of radix-64 encoding.

---

[12] https://www.openmicroscopy.org/site

[13] http://www.openmicroscopy.org/site/products/bio-formats

[14] https://www.openmicroscopy.org/site/support/ome-model/ome-tiff/index.html

Bio-Formats library integration
The Bio-Formats library has a standardized API which is integrated into open-source analysis programs like ImageJ, CellProfiler, and Icy, commercial scripting environment like MATLAB, and the web image processing pipeline (WIPP). The Bio-Formats library is also a key component in the image database solutions like OMERO and the JCB DataViewer (used by the *Journal of Cell Biology* for hosting data associated with submissions). WIPP leverages the Bio-Formats and OME-XML Java libraries for server side image reading and writing. The Java program called ConvertToOmeTiff.java is provided by OME and is invoked when files are uploaded. This converts each file into OME-TIFF format. Conversions can be replicated by following the instructions online.[15]

5.2.5   Basic Image Processing Using ImageJ/Fiji

Among all image processing implementations in Table 5.2, we chose ImageJ/Fiji to illustrate its use for basic image processing. Our selection is based on the popularity of ImageJ/Fiji for bioimage analysis [8]. The software is completely open source and has a free license, it has multiple modes of usage (see Table 5.2), and its JVM-based implementation makes it operating system (OS) independent. ImageJ also allows us to demonstrate successful modes of research involving some effort using GUI interfaces, macro languages, scripting languages, library APIs, and workflow systems.

Using ImageJ/Fiji
To learn more about ImageJ, we will refer the reader to the publication entitled "Analyzing fluorescence microscopy images with ImageJ" by Peter Bankhead [8]. The technical report introduces basic image processing and the implementations in ImageJ and Fiji. It then covers general methods such as Gaussian filters, thresholding, spot detection, and object analysis, with some macro-writing exercises to extract information from many types of microscopy images. The technical report ends with a focus on the specifics of extracting and interpreting measurements from fluorescence images.

Let us illustrate the use of ImageJ for three different tasks:

1. Subtracting a mean-filtered image from itself
2. Performing the same operation on a collection of images
3. Processing an image that is larger than the memory of your computer

Task 1 (Subtracting a mean-filtered image from itself) The first task can be completed using the GUI interface by following the steps below:

Step 1: Install ImageJ
Step 2: File→Open (load an image img_r005_c002.tif from the small test dataset):

Width: 851.9045 µm (1392 pix)
Height: 636.4804 µm (1040 pix)
Size: 2.8 MB

---

[15] https://www.openmicroscopy.org/site/support/ome-model/ome-tiff/code.html

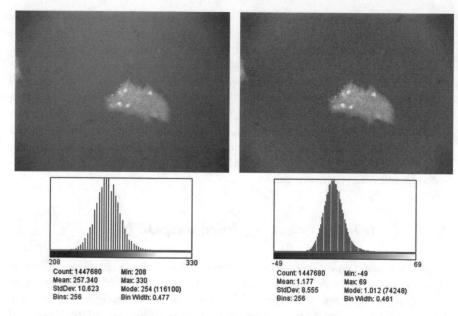

Fig. 5.1 Comparison of histograms before and after subtracting mean-filtered image from itself. Left – original image. Right – mean-subtracted image

Step 3: Image→Duplicate (create a copy of img_r005_c002.tif).
Step 4: Filters→Mean→radius = 500 (create filtered image).
Step 5: File→Save as TIFF (save result as img_r005_c002_mean500.tif).
Step 6: Process→ImageCalculator → set:

   image1: img_r005_c002.tif
   Operator: Subtract
   Image 2: img_r005_c002_mean500.tif
   Check box: 32-bit (float) result

In Fig. 5.1, we can see from the image histograms that the mean value has shifted from 257.34 before processing to 1.177 after processing.

Task 2 (Subtracting a mean-filtered image from itself over an image collection) The second task can be performed using scripting since the GUI interface would require 33 interactions to apply the same sequence on the image collection in "Small_ Fluorescent_Test_Dataset." The easiest method for creating a script is to open the recorder from Plugins→Macros→Record, record the sequence of operations, and then create the script file by clicking on "Create" button in the recorder. In this case, the script in ImageJ macro language would be:

```
open("C:\\chapter5\\img_r005_c002.tif");
run("Duplicate...", " ");
run("Mean...", "radius=500");
```

```
imageCalculator("Subtract create 32-bit", "img_r005_c002.tif","img_
r005_c002-1.tif");
selectWindow("Result of img_r005_c002.tif");
```

To extend the script to work over all files in a folder, the script must become more generic. One way to expand the script is to invoke a directory chooser dialog and then loop over a file list. This approach is illustrated in the script below. The command setBatchMode(false) is used to view the windows with results; otherwise they would be suppressed. The test of TIF suffix is an added filter to choose only the files of interest.

```
dir1 = getDirectory("Choose  Directory ");
list = getFileList(dir1);
setBatchMode(false);
for (i=0;i<list.length;i++) {
    if (endsWith(list[i],"tif") || endsWith(list[i],"TIF")){
        open(dir1+list[i]);
        run("Duplicate...", " ");
        nameDup = getTitle();
        run("Mean...", "radius=500");
        imageCalculator("Subtract create 32-bit", list[i],nameDup);
        selectWindow(getTitle());
    }
}
```

Task 3 (Processing an image that is larger than the memory of your computer)   The third task is to process an image that is larger than the available computer memory. In this case, we need to work with two image instances during mean filtering (original and filtered) and three image instances during subtraction. The resulting image has 32 bits per pixel (BPP) as opposed to the inputs with 8 BPP which has implications for memory (RAM) usage. We have also noticed that computing the mean-filtered image with a very large kernel size (i.e., radius = 500) for a 1.3 Megapixel image can be slow on a laptop. ImageJ could be reinstalled on a machine with more RAM or the software could be rewritten. Similarly, the speed of execution could be improved by running ImageJ on a machine with a faster processor or by redesigning the software to utilize existing hardware more efficiently. This dynamic interplay between advanced hardware capabilities, increasing size of image data, and the cost of redesigning legacy software leads to community efforts such as ImageJ2[16] (software auto updater, enhanced file format support, additional support of several scripting languages, etc.). The WIPP system is another approach on how to address the dynamics between hardware, data, and software. Chapter 6 is devoted to the interplay between hardware capabilities and software design.

---

[16] imagej.net/ImageJ2

5.3   Overview of Algorithms in WIPP

The algorithms in a web image processing pipeline are listed as computational jobs. Every algorithm takes input data and generates output data. The types of data and computational jobs are provided in Table 5.3. The characteristics of algorithms behind each computational job are described next.

Characteristics of algorithmic computations

The algorithms run either on a server or in the web browser. Each algorithm also has input parameters, an explanation of the parameters, and its own web page with a user interface for configuring inputs before the computational job is launched. Table 5.4 summarizes computational types of algorithms, their execution location, input and output type, and a list of parameters. To discriminate types of image collections, we introduce the following hierarchical terms describing the image content:

- *Raw collection* is a set of intensity images as acquired by a microscope:
  - Grid collection is a subcategory of raw collection where the microscope acquisition forms a grid of images.
  - Dark collection contains an image acquired by closing a camera aperture.
  - Fluorescein collection contains an image acquired with fluorescent media but without cells.

- *Binary collection* contains images with two unique intensity values denoting foreground and background (e.g., foreground is set to 255 and background to 0).
- *Labeled collection* is a set of segmented images with a unique label/intensity assigned to each contiguous region of interest.

    These terms are used in Table 5.4 as well as in the WIPP user interface.
    We will describe a subset of the algorithms in the next sections. The algorithmic descriptions are presented in the order that they are used in a typical image processing

Table 5.3 A list of data and computational jobs accessible via web user interfaces

Data	Computational jobs
Tile collections	Flat-field correction
Stitching vectors	Background correction
Tracking vectors	Filtering
Image pyramids	Stitching
Visualizations	EGT segmentation
	FogBank segmentation
	Mask labeling
	Pyramid building
	Image assembling
	Intensity scaling
	Tessellation

Table 5.4 Description of algorithmic computations and input/output types

Computation type	Computation location	Input type	Parameter list	Output type
Spatial filter: Mean, median, Gaussian, morphological erosion or dilation	Server	Raw collection	Kernel size	Collection
Segmentation: Empirical gradient threshold method	Server	Raw collection	Min object size, max hole size, threshold adjustment delta	Collection
Segmentation: FogBank method	Server	Collection (raw + labeled)	Border mask percentile threshold, minimum seed size, minimum object size, direction	Collection
Mask labeling	Server	Binary collection	Pixel connectedness	Collection
Calibration: Flat-field correction	Server	Collections (raw + dark + fluorescein)		Collection
Calibration: Background correction	Server	Collections (raw + binary)	Gap size, ring size	Collection
Stitching	Server	Grid collection	Stitching options	Stitching vector
Tracking	Server	Labeled collection (one image per time frame)	Minimum object size, maximum centroid displacement, enable cell division and/or fusion	Collection + tracking vector
Image assembly	Server	Grid collection Stitching vector		Collection (one image)
Intensity scaling to 8 bits per pixel	Server	Raw collection	Range min and max	Collection
Multi-resolution pyramid building	Server	Grid collection Stitching vector		Pyramid
Mask generator by tessellation	Server		Tile size and shape, image mask size	Collection (one image)
Deep zoom configuration	Client	Set of pyramids	Definition of layers and measurement widgets	Deep zoom visualization
Spatial filters: Sobel edge, morphological erosion or dilation	Client	Pyramid tiles	Kernel size	Deep zoom
Intensity filters: Threshold, contrast, gamma, brightness, saturation, hue, etc.	Client	Pyramid tiles	Parameter	Deep zoom
Connectivity analysis and region area	Client	Canvas		Image + table

workflow designed to extract object measurements. They are divided into the following categories:

1. Image correction
2. Stitching and mosaicking
3. Object segmentation, tracking, and feature extraction
4. Intensity scaling and image pyramid building

5.4 Image Correction Algorithms

Microscopy images are acquired as a 2D matrix of intensity measurements. However, the magnitude of each intensity is affected not only by the imaged specimen but also by the imaging optics, detector, illumination source, and the geometry of these three components. Thus, image intensities must be corrected for these spatially varying effects. We will describe dark current, flat-field, background, and noise corrections.

5.4.1 Dark Current Correction

To perform the dark current correction, an image is acquired with a closed camera shutter. This image is an estimate of the dark current due to thermal fluctuation of the charge-coupled device (CCD) used as a detector. Dark current correction is performed by subtracting the "dark current" image from each acquired image.

Note To obtain a representative dark image, we recommend acquiring multiple dark images (more than five) and then computing the median image from the multiple acquired images.

5.4.2 Flat-Field Correction

The flat-field correction is needed to correct for the vignetting effect (the darkening of image corners relative to the center). Unlike the additive dark current effect, vignetting is a multiplicative effect. Thus, a flat-field correction model includes a division and is given by the equations below:

$$I_F^t = \frac{a}{b} \tag{5.1}$$

$$a = I^t - D \tag{5.2}$$

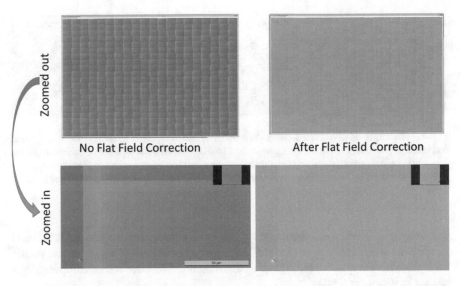

Fig. 5.2 Examples of stitched frame without flat-field correction (left) and after correction (right). The bottom rows show the zoomed version of the top row

$$b = \frac{(F - D)}{\max(F - D)} \tag{5.3}$$

where I_F^t is the flat-field corrected image at time frame t, I^t is the acquired image at frame time t, D is the dark current image, and F is the fluorescein image (an image acquired with fluorescent media but without cells). F and D are independent from time during an acquisition. Figure 5.2 illustrates the visual effect of flat-field correction over a large image frame consisting of many fields of view. Notice the vignetting effect in the image on the top left of Fig. 5.2 with brightness discontinuities between adjacent fields of view. After flat-field correction, the brightness discontinuities are minimized as illustrated in the top right image of Fig. 5.2.

Note Other methods have been used in practice. For example, one could estimate the flat-field image by applying a Gaussian filter with a very large kernel size and then dividing the measured image by the filtered image to obtain the corrected image.

5.4.3   Background Correction

The background is an additive signal present in most microscopy images. The background intensity comes from the media surrounding the cells. The pixel intensity value of the background is unknown underneath the regions of interest (e.g., cells or

Fig. 5.3 Example of uncorrected (left), dark and flat-field corrected (middle), and background corrected (right) images. The images correspond to a mosaic of 22 × 18 FOVs acquired from the green fluorescent protein (GFP) image channel of stem cells on their fourth day from seeding. The five red "x" marks in the middle image correspond to FOVs with nothing but background pixels in them

cell colonies). This signal should be subtracted from the acquired image intensities in order to estimate the pure signal S^t for a given time point. This operation should take place before making biological measurements. The pure signal computation is shown in Equation ((5.4)):

$$S^t = I_F^t - B \tag{5.4}$$

where S^t is the pure signal and B is the background image estimated after flat-field and dark current corrections. Our goal is to estimate the background intensity from the neighboring background pixels of the regions of interest.

Application to fluorescent images
Background correction is needed in fluorescent imaging due to the complex interaction of foreground objects and their surroundings. For example, foreground objects such as living cells are fluorescently labeled with a marker and surrounded by media. The intensity at a cell location contains contributions from non-specific autofluorescence of the cell culture media, culture dish, and any extracellular matrix protein coatings. The fluorescent components may vary spatially and temporally because of the distribution of fluorescent molecules, different amounts of light interacting with these molecules, and photo-bleaching of molecules over time. It is impossible to capture these spatiotemporal variations of fluorescent signal in the background from a single FOV since stem cell colonies grow to cover very large spatial areas. Figure 5.3 illustrates the impact of dark current, flat-field, and background correction on the resulting image. One of the main challenges is the temporal background correction as illustrated in Fig. 5.4. In this experiment, the culture media for living cells needed to be changed every 24 h which caused significant background intensity variations over time. The curves illustrate global intensity variations across an entire image mosaic and over time.

Background estimation algorithm
We will overview two approaches to background correction that can take advantage of large image experiments [9]. Both approaches assume that the background pixels form a continuous surface within a FOV and across multiple spatially overlapping

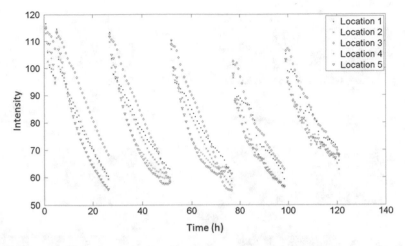

Fig. 5.4 Spatiotemporal graphs of average intensity per FOV: These results are displayed for the five numbered locations shown in Fig. 5.3 through 5 days of acquisition

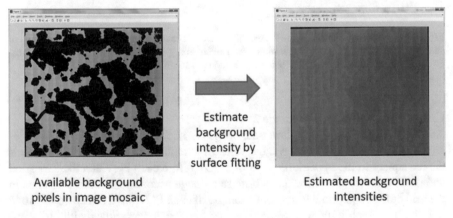

Available background pixels in image mosaic

Estimate background intensity by surface fitting

Estimated background intensities

Fig. 5.5 Background intensity estimation by surface model fitting to an image mosaic based on available background pixels (nonblack pixels in the image on the left)

FOVs. Thus, the background intensities can be estimated from one of the following pixels:

1. A subset of pixels in a stitched FOV (denoted as "surface fit of entire image")
2. Pixels at the same locations in multiple FOVs (denoted as "creation of sub-mosaics")

Surface fit of entire image (scheme 1)

One option is to detect background pixels at each time frame and then use surface fitting to estimate background intensities at every pixel location. This approach is illustrated in Fig. 5.5.

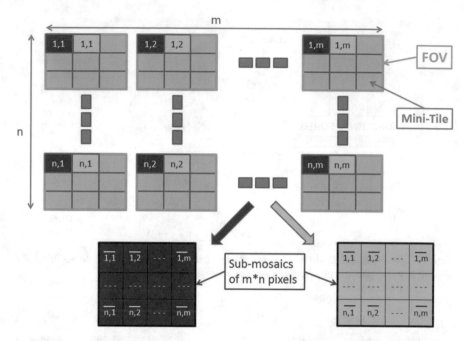

Fig. 5.6 Image partitioning Scheme 2: Average values of the red pixels from the same location of all FOV are assembled into the red sub-mosaic. The red and green colors denote the original location of the pixels in each FOV and their new locations in sub-mosaic images

Creation of sub mosaics (scheme 2)

Another option is to create sub-mosaic images as shown in Fig. 5.6. The subimages are created by extracting the average value of a subtile (an example of a subtile is the red square in the upper left corner of a FOV) from each FOV at a fixed location and placing it into a constructed sub-mosaic image according to the FOV index in the grid of FOVs per time frame. For example, if each FOV has a dimension of 1040 × 1392, then each FOV will be tiled into 65 × 87 = 5655 subtiles with (16 × 16) pixels per subtile. Each subtile at a particular location will be replaced by the average value over (16 × 16) pixels. These values from the same location across all FOVs (e.g., the red) will be assembled together to form one sub-mosaic. Each sub-mosaic will have a size of (16 × 22) pixels. The total number of sub-mosaics formed is 5655. The surface fit will be applied on each mosaic independently. Figure 5.7 shows the sub-mosaic images formed from the (16 × 22) grid of FOVs. The black pixels indicate that they belong to the foreground and hence are not available for background modeling. The idea behind the sub-mosaic creation is that background intensities are continuous throughout the entire plate and across FOV boundaries. The sub-mosaic image is a good approximate map of the background throughout the entire mosaic given a FOV location.

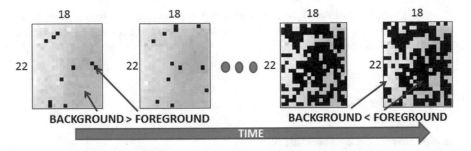

Fig. 5.7 Resulting sub-mosaic images over time: The decreasing number of background pixels is due to the growth of stem cell colonies

Metrics for image correction evaluation

Microscope image corrections can be evaluated in terms of:

- Root mean square (RMS) of the difference between the modeled background intensities and the observed intensities over the background reference pixels (BRP)
- Signal-to-noise ratio (SNR)

RMS and SNR errors of a given image are computed according to the equations below:

$$SNR = \frac{1}{N_{\mathrm{FRG}}} \sum_{n \in \mathrm{FRG}} \left(I_n - \hat{B}_n \right) \Big/ \mathrm{std}\left(\left\{ I_n - \hat{B}_n \right\}_{n \in \mathrm{BKG}} \right) \tag{5.5}$$

$$\mathrm{RMS} = \sqrt{\frac{1}{N_{\mathrm{BRP}}} \sum_{n \in \mathrm{BRP}} \left(I_n - \hat{B}_n \right)^2} \tag{5.6}$$

where N_{FRG} and N_{BRP} denote the number of foreground pixels (FRG) and BRP, respectively, n is an index for pixel location in the image, I_n is the observed intensity of pixel n, B_n is the modeled background intensity of pixel n, FRG is the set of pixel locations that are identified as foreground, BKG is the complement of FRG and contains the set of pixels identified as background, and BRP is a subset of BKG used for background model assessment. The optimal correction will minimize RMS and maximize SNR. The dark current noise can be used as a benchmark for the lowest noise level that an imaging system can achieve after all corrections.

Notes about the two schemes

Based on the RMS and SNR metrics, Schema 2 (creation of sub-mosaics) outperforms Schema 1 (surface fit of entire image) [9]. The background correction following Schema 1 is available in WIPP. The other, Schema 2 with sub-mosaics, has not been added to WIPP yet, but it is suitable for parallel execution on computational resources with limited RAM. It should be noted that the aforementioned image corrections still do not account for photo-bleaching effects, movement of the media, and illumination variation.

5.4.4   Noise Filtering

Image correction by noise filtering

In the previous considerations of image corrections, we ignored additive noise ε_{ijt}. In many microscopes, there may be a priori knowledge about the model of additive noise. A noise filtering operation can help to remove the noise (called image denoising). However, denoising is a difficult problem because we do not know the boundary between the noise and the signal (noise is a random process). We also do not know which method for noise removal would be theoretically the best. In other words, we need to decide on the type of the spatially varying filter and its spatial extent. Most denoising approaches are based on suppressing very high-frequency and incoherent components of an input image.

Noise types

The two most frequently assumed noise models in microscopy are additive white Gaussian noise (AWGN) and Poisson noise. The models for their probability distribution functions $P(I)$ are given by

$$P^{Gaussian}\left(I\right)=\frac{1}{\sqrt{2\pi\sigma}}e^{-\frac{\left(I-\bar{I}\right)^2}{2\sigma^2}} \tag{5.7}$$

$$P^{Poisson}\left(I^{RAW}=I\right)=\frac{\lambda^I}{I!}e^{-\lambda} \tag{5.8}$$

where I is the image intensity at (i,j) and takes discrete values, $\bar{I}$ is the average intensity, σ is the standard deviation of all intensity values, and λ is the average number of shot noise events per interval. Shot noise events occur due to random fluctuations of the electric current (a flow of electrons) in microscope electronic circuits.

Filter type for a noise model

Empirical studies have been performed to determine which type of spatial filter is optimal given a noise model. According to [10] a median filter is best for Gaussian and Poisson PDF models when evaluated using the following peak signal-to-noise ratio (PSNR) definition:

$$PSNR=10\log_{10}\frac{V^2}{\sum_i^M\sum_j^N\left(I^{After}\left(i,j\right)-I^{Before}\left(i,j\right)\right)^2} \tag{5.9}$$

Table 5.5 Filter kernel of
size 3×3 with nine
coefficients

w_1	w_2	w_3
w_4	w_5	w_6
w_7	w_8	w_9

For PSNR, V is the maximum value for the discrete intensity range, and I^{After} and I^{Before} are the intensities before and after applying the filtering operations.

For the purposes of noise removal, we classify filters based on frequency range and linearity.

Frequency-based classification
A spatial filter is characterized by its 2D matrix of coefficients called a kernel (see Table 5.5). Filters are classified as low-pass, high-pass, band-pass, or band-reject. For instance, a mean filter with all $w_k = 1$ preserves low frequencies and hence is a low-pass filter.

Linearity-based classification
Another filter classification is into linear and nonlinear categories. A linear filtering method would be described by a linear weighted combination of intensity values (i.e., the filtered value is a weighted sum of the input pixels). For instance, a convolution of a spatial filter with an image is linear filtering as documented in the equation below:

$$I^{\text{Filtered}}(i,j) = w(i,j)^* I(i,j) = \frac{\left| \sum_{r=-n/2}^{m/2} \sum_{c=-n/2}^{n/2} \left(w(r,c) I(i-r,j-c) \right) \right|}{\sum_r^m \sum_c^n w(r,c)} \tag{5.10}$$

where * denotes the convolution operation, $w(r,c)$ is the coefficient of the kernel of size $m \times n$ at location (r,c) inside of the kernel, and $I(i-r,j-c)$ is the image intensity in the coordinate system of the kernel. The denominator is often set to one by design. Nonlinear filtering methods cannot be written as a linear combination of intensity values. For example, min, max, or median filters are nonlinear filters, which can be referred to as order statistics or rank filters. The equation below illustrates the computation of minimum filtered image:

$$I^{\min\_Filtered}(i,j) = \min_{(r,c) \in \left[-\frac{m}{2}, \frac{m}{2} \right] x \left[-\frac{n}{2}, \frac{n}{2} \right]} \left\{ I(i-r,j-c) \right\} \tag{5.11}$$

Filters in WIPP
WIPP offers mean, median, min, max, and Gaussian blur filters. The filters are defined over a region using a radius parameter. The radius refers to the kernel size to $(2 \times \text{radius} + 1) \times (2 \times \text{radius} + 1)$ instead of $m \times n$ for mean, min, and max filters as illustrated in Fig. 5.8 for the case of min filtering. The radius parameter of the Gaussian blur filter refers to the standard deviation σ, and the kernel size is based on the non-zero discrete values of the Gaussian PDF.

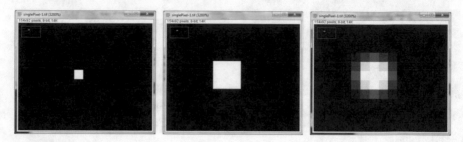

Fig. 5.8 Filtering a single bright pixel (*left*) by a maximum filter (*middle*) and by a Gaussian blur filter with the radius parameter set to 1

Table 5.6 Example of a Gaussian blur kernel of size 5×5 derived for $\sigma = 1$

1	3	5	3	1
3	15	25	15	3
5	25	41	25	5
3	15	25	15	3
1	3	5	3	1

The radius value is equal to 2

For example, in both the ImageJ and WIPP implementations, the kernel for the Gaussian blur filter with $\sigma = 1$ is approximated by a 5×5 kernel with the coefficients shown in Table 5.6. For computational efficiency, the kernel is truncated to 5×5 instead of 7×7. Another note about these implementations is that the normalizing multiplier is set to $\frac{1}{255}$ instead of $\frac{1}{249}$ which would be the sum of all coefficients.

There are many more image correction models beyond those available in WIPP. For example, if out-of-focus images are of concern, then one could devise a deconvolution method to remove the blur due to a point spread function and the inaccuracy of manual focusing. As with all image corrections, the estimation of parameters is critical to improving the quality of image-based measurements. Another important factor is the order in which image correction operations are applied to raw images. Modeling, model parameter estimation, and processing order are still open research topics in microscopy image analyses.

5.5 Algorithms for Stitching and Mosaicking Many Images

In this section, we motivate acquisition of many fields of view (FOVs) with or without spatial overlap. FOVs acquired with spatial overlap can be put together by using a stitching algorithm, while FOVs acquired without spatial overlap can be assembled by a mosaicking algorithm. We focus on stitching and mosaicking algorithms as implemented in WIPP.

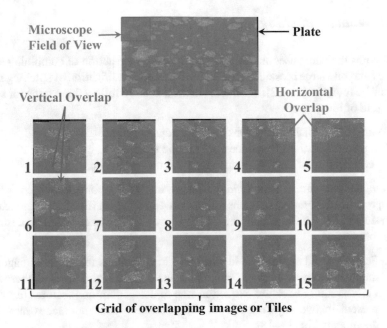

Fig. 5.9 Grid of overlapping image tiles

Motivation behind many FOVs

Cell culture microscopy must address the spatial scale mismatch between the microscope's FOV needed for observing an object of interest and the much larger size of the studied specimen containing many objects of interest. This can be illustrated by considering the area of a standard 6-well plate well that is approximately 1000 times larger than the FOV acquired with a 10× objective. Automated microscopy overcomes this issue by acquiring a grid of partially overlapping images (tiles) that cover most of the experimental area (Fig. 5.9). Better statistical sampling and the capture of rare dynamic events motivate large coverage imaging.

The difference between stitching and mosaicking

In many applications, one large FOV is imaged as a collection of small FOV. The goal of stitching is to estimate spatial position coordinates for each small FOV in a Cartesian coordinate system of the large FOV. The goal of mosaicking is to assemble small FOVs into a large FOV image. Mosaicking consists of compositing small FOVs and blending spatially overlapping regions. The motivations for separating these two operations are that:

- Coordinates originate not only from estimation methods but also from devices and users.
- Mosaicking may be too memory intensive for a regular desktop and therefore should not be integrated with stitching.
- Stitching and mosaicking have their own independent sets of parameters.

We describe stitching and mosaicking algorithms next.

5.5.1   *Image Stitching*

Stitching is the name used in literature that refers to the action of combining single images into one large mosaic. A detailed survey and classification of stitching methods can be found in [11, 12]. Most tools in the literature follow three steps for stitching a grid of image tiles:

1. Compute candidate translations between adjacent tiles.
2. Optimize translations to reduce errors in the stitched image.
3. Order assembly of tiles to produce the final mosaic image.

1. *Compute candidate translations*: There are two commonly used general approaches to compute the translations between adjacent tiles: (1) image feature-based and (2) Fourier transform-based. These two approaches can be described as follows:

   1. Feature-based approaches identify matching features in adjacent images and then use these features to compute image translations. However, these techniques require a feature extraction step to detect common features of interest present in two adjacent images. The feature-based approaches have been documented in [13–18].
   2. Approaches based on Fourier transforms use image frequency components as the main features. This approach assumes that images have enough pixels with unique frequency components in the overlapping image areas. The Fourier transform-based approaches have been published in [19–25].

2. *Optimize translations to reduce errors*: Many techniques can be used to optimize computed translations and reduce the errors in the stitched image:

   • Weighted least squares can be applied to all translations maximizing the weights of features found in the overlap areas.
   • Minimize an error function between hypothesized and actual point correspondences using a joint registration algorithm.
   • Maximize the normalized cross-correlation.
   • Maximize/minimize a predefined energy function.
   • Use global geometric and radiometric parameter estimation.

3. *Order assembly of tiles into a mosaic image*: Multiple approaches exist to assemble the mosaic based on the computed and optimized translations. Some treat the translations and their corresponding normalized translation weights/confidence as an undirected graph where an algorithm like the minimum spanning tree is used to assemble the mosaic image. Others treat the problem as an over-constrained system of linear equations and solve it using least-squares methods, iterative square displacement minimization, or singular value decomposition. The choice of which approach to use for any step depends on the overall image content and the characteristics of matching features.

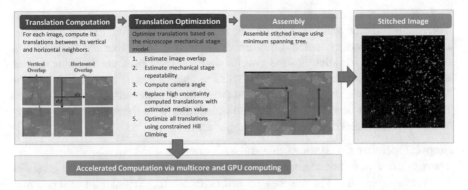

Fig. 5.10 Overview of MIST stitching algorithm

Introduction to Microscopy Image Stitching Tool (MIST)
WIPP uses the MIST algorithm [26] to execute the three steps of stitching and generate a stitching vector. The stitching vector consists of a list of file names and their positions in a Cartesian coordinate system of the mosaic image (i.e., large FOV). The final composition into a mosaic image is implemented as a separate job in WIPP called "image assembly."

Figure 5.10 shows a top-level stitching overview with the translation estimation followed by translation optimization per FOV and ordering of tile assembly in the left three blocks. The stitching accuracy is achieved by estimating mechanical stage model parameters from computed pairwise translations. The stage parameter estimates minimize stitching errors by constraining and optimizing the translations within a square area four times the size of the microscope stage repeatability. The stage repeatability is the ability of the microscope stage to come back to the same location after visiting other locations. Meaning that if the user asks the stage to go to location (x_1, y_1), after visiting other locations, the stage will come back to a neighboring location of $(x_1 \pm r, y_1 \pm r)$. Based on our experience, y a good light microscopy automated stage has a repeatability of $r = 1$ μm. The MIST implementation has been optimized for speed and memory efficiency allowing it to be used on large image collections [27]. The stage modeling and error minimization methodology is applicable to microscopy images or any mechanical instrument that acquires a 2D grid of image tiles.

Next, we will describe the details of the three steps (translation computation, optimization, and order of tile assembly) as implemented in the MIST algorithm.

Step 1: Compute candidate translations
MIST uses a Fourier-based approach because of its simplicity and predictable performance. One of the advantages is that this approach does not need an additional feature detection tool. MIST implements the phase correlation method (PCM) to compute translations between adjacent tiles. PCM is based on the Fourier shift theorem which computes the spatial shift between two images as a phase shift in the frequency domain.

In real images, phase correlation (PC) contains several peaks that correspond to different translation values. To determine the correct translation, the top two peaks in the PC matrix are evaluated. The number of peaks is adjustable and the default is two. Due to the periodicity of the Fourier domain, each peak corresponds to four different possible translations (in 2D). MIST evaluates these four possible translations, for each peak, using the normalized cross-correlation (ncc) of the overlap area between adjacent images. The candidate translation with the highest ncc value is selected as the translation between two adjacent images.

Step 2: Optimize translations to reduce errors
There is a degree of uncertainty in the translation computation that causes errors in the stitching results. The sources of errors that affect translation computation between pairs of images are the:

1. Signal-to-noise ratio in the acquired image
2. Amount of signal in the overlap area
3. Signal distribution with respect to the stage movement in the overlap area (i.e., flat-field effect in some imaging modalities, uniform or periodic signal distribution)
4. Mechanical imperfections of the automated microscope stage (i.e., stage repeatability and actuator backlash)

Moreover, the mechanical stage model parameters vary over time, and the magnitude of variation depends on the microscope usage. If the user of a microscope has calibrated the equipment and has measured the stage repeatability, then there is an option to specify such parameters in the tool. If the user has provided those parameters, then MIST does not estimate them. However, it is difficult and time-consuming to calibrate and estimate the mechanical stage properties of the microscope.

Additionally, there are research environments where a microscope might have multiple users. Each user might adjust/change/perturb the camera settings or mechanical stage. These physical changes can alter the microscope configuration to be out of calibration. Finally, there are time-lapse experiments in which touching the mechanical stage cannot be avoided. For example, cells in live experiments need to be fed intermittently. To do this, the sample is removed from the microscope and then put back on the stage after media exchange. This unavoidable feeding operation can alter the mechanical stage properties among many other experimental settings.

To address the aforementioned challenges, MIST offers an automated way to estimate the stage parameters from the computed translations to prevent the user from having to calibrate and measure these parameters before every acquisition. We outline the method for estimating mechanical stage model parameters and then use the estimates for constrained translation optimization.

Estimation of mechanical stage model from pairwise translations
An automated microscope has two coplanar coordinate frames, the observation frame (i.e., camera) and the control frame (i.e., stage actuators). They are related by the camera angle α, as shown in Fig. 5.11 (a). This angle is difficult to calibrate.

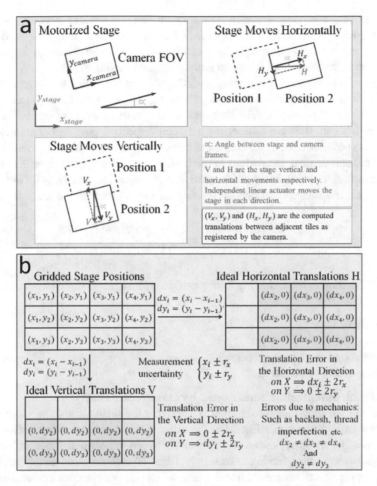

Fig. 5.11 Stage mechanical model. (**a**) Stage displacements as observed by the camera. (**b**) Uncertainty and errors of horizontal and vertical tile translations due to stage mechanical properties

Therefore, a misalignment between the camera and stage axes will remain in most experiments. The camera observes the horizontal and vertical stage movements, H and V, as (H_x, H_y) and (V_x, V_y) which are computed as follows:

$$\begin{bmatrix} H_x \\ H_y \end{bmatrix} = \begin{bmatrix} H\cos\alpha \\ -H\sin\alpha \end{bmatrix} \& \begin{bmatrix} V_x \\ V_y \end{bmatrix} = \begin{bmatrix} V\sin\alpha \\ V\cos\alpha \end{bmatrix} \tag{5.12}$$

A motorized mechanical XY stage moves a biological sample relative to the microscope's optical column. This movement is carried out by two independent stepper motor linear actuators, one for each direction. The mechanical uncertainty of such a system is known as the stage repeatability. Moreover, the imperfection of the stage as a mechanical device introduces a variable overlap between adjacent tiles. Modeling the mechanical properties of a stage provides an upper bound to the variable overlap between images and can be used to limit the search for optimal translations, thereby minimizing the margin of stitching error.

Figure 5.11 (b) shows a grid tiling with the positions (x, y) that the stage will visit. Each position has an uncertainty equal to the stage repeatability $(x \pm r_x, y \pm r_y)$. However, translations (dx, dy) computed in the vertical or horizontal directions between adjacent tiles are differences between respective positions. Therefore, the maximum possible error in the computed translation values is $(dx \pm 2r_x, dy \pm 2r_y)$. The horizontal and vertical translations in the image coordinate system must account for the camera angle as well as the mechanical uncertainties. The equations for horizontal and vertical translations that include the microscope models are the following:

$$\begin{cases} H_x = dx_i \cos(\alpha) \pm 2r_x \\ H_y = dx_i \sin(\alpha) \pm 2r_y \end{cases} \text{and} \begin{cases} V_x = dy_i \sin(\alpha) \pm 2r_x \\ V_y = dy_i \cos(\alpha) \pm 2r_y \end{cases} \tag{5.13}$$

MIST estimates the following four quantities from the translation matrices H and V:

1. Overlap amount
2. Camera angle
3. Microscope stage repeatability
4. The microscope backlash

Translation optimization constrained by stage repeatability

Figure 5.11 (b) shows each column in H as having the same dx_i and the same dy_i for each row in V within a $\pm 2r$ limit. On the other hand, dx_i values differ between the columns of H, while dy_i values are different between the rows of V due to backlash and mechanical imperfections. Thus, we filter H column-wise and V row-wise where we replace all computed translations, whose dx_i or dy_i values deviate from the median value by more than $4 \times r$ (the stage repeatability), by the median value in that direction. We then apply constrained hill climbing to the ncc values centered at the median translation and constrained within $4r \times 4r$ region. Hill climbing will find the translation with the maximum ncc value by following the steepest gradient. The $4r \times 4r$ constrain comes from the model and bounds the algorithm to a small search space while converging to an optimal value.

Step 3: Order assembly of tiles to form a stitching vector

This assembly problem can be represented as an undirected graph where vertices are tiles and normalized correlation coefficients are edges. Each tile is connected to its surrounding four neighbors, three neighbors for tiles on edges, and two neighbors for tiles on corners. This over-constrained problem needs to be resolved to construct a well-formed image. We use the weighted maximum spanning tree algorithm to find the optimal subset of edges. The edges connect all tiles together without any circular subsets of edges per tile (each tile is connected only once to the reconstructed image). The weighted maximum spanning tree algorithm maximizes the sum of all weights along the path defined by all edges. The weights of all computed translations that satisfy the physically plausible offset stage model criteria (offsets <4r) are increased to ensure preferential selection of these translations during assembly.

5.5.2   Image Mosaicking

A large mosaic image can be assembled from an image collection based on position information about each FOV image by using the image assembly job. The position information can come from one of the following inputs:

(a) A stitching vector
(b) A metadata file with coordinates of a motorized stage
(c) The user specifications via file naming pattern

To generate the vector with position information, run the stitching job and specify one of the four options:

1. Stitching (MIST algorithm generates a stitching vector)
2. TIFF stage metadata (WIPP extracts a stitching vector based on metadata in acquired image tiles)
3. No overlap mosaic (WIPP generates a stitching vector that defines positions based on a spatial file name pattern without any spatial overlap)
4. Time sequence of one FOV (WIPP derives positions based on a temporal file name pattern only)

The image assembly job takes the image collection and a stitching vector to produce a composed mosaic image. For mosaic images that fit to computer memory, there are two cases for image assembly:

1. With spatial overlap
2. Without spatial overlap

Mosaic assembly with spatial overlap
In this case, compositing is performed by using the stitching vectors generated by the MIST stitching algorithm or derived from TIFF stage metadata. For the pixels in the overlapping regions, intensities are computed using a linearly weighted blending. Blending compensates for illumination differences in overlapping areas of spatially adjacent tiles. However, blending will not correct for spatial misalignment that can be significant if a microscope has not been properly calibrated. Ideally, scientists would like to have access to all values at a given spatial location that were processed during blending.

Mosaic assembly without image overlap
In this case, compositing is executed using the stitching vectors derived only from spatial or temporal file name patterns ("no overlap" and "time sequence of one FOV" options in the stitching job). No image blending is performed since image tiles are just placed side by side. The resulting mosaic file can be downloaded for viewing or used for object feature extraction.

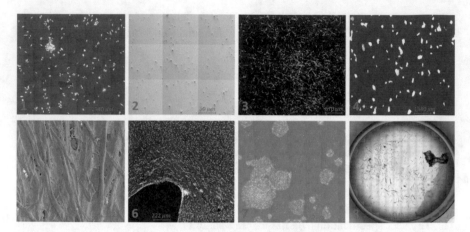

Fig. 5.12 Stitching application examples: (1) A10 cells, (2) carbon nanotubes, (3) HBMSC, (4) IPS cell colonies, (5) paper nanoparticle, (6) rat brain cells, (7) stem cell colonies, and (8) worms

For mosaic images that do not fit in computer memory (i.e., image assembly job fails due to an out-of-memory error), we recommend running the pyramid building job with the stitching vector as input and then view the image content in WIPP. The pyramid building job will convert small FOVs into a Deep Zoom viewable representation without compositing the large FOV image.

5.5.3 Practical Remarks

We overviewed the MIST stitching algorithm and a few mosaicking options. The stitching algorithm has been tested on a variety of image collections acquired using five microscopes (Leica, Olympus, Nikon, Zeiss, and FIB SEM (focused ion beam scanning electron microscope)) and in four imaging modalities (fluorescent, phase contrast, bright-field, and FIB SEM). The tests included diverse image content, overlap values ranging between 10 % and 70 %, and grid sizes ranging from 5 × 5 to 70 × 93. Figure 5.12 shows examples of some of the image content like A10 cells, brain cells, carbon nanotubes, human bone marrow stromal cells (HBMSC), IPS cells, paper sample, stem cells, and worms.

However, there are image sets for which stitching can fail due to low SNR and/ or a lack of image features. For some of these cases, the stitching algorithm can be assisted by stitching the filtered and segmented images rather than the raw images.

Mosaicking images is executed by running the stitching job followed by the image assembly job. The stitching job generates the vector with position information for each small FOV in the coordinate system of the large FOV. If the mosaic image is larger than computer memory, then pyramid building should be executed instead of image assembly to view the mosaic image in WIPP.

5.6   Object Segmentation, Tracking, and Feature Extraction Algorithms

To quantify objects of interests from time-lapse large FOV images, algorithms for the following three applications must be designed:

1. Spatial segmentation of objects
2. Temporal association of spatial segments
3. Extraction of spatiotemporal characteristics of objects

The objects of interest can be cells, cell colonies, or other objects. Next, we introduce readers to algorithms that have been designed for performing each of these tasks.

5.6.1   Object Segmentation

Segmentation introduction

Segmentation is one of the fundamental digital image processing operations [28]. It is used across all scientific fields where imaging is a quantitative measurement method. In computer vision, image segmentation is the process of partitioning an image into multiple segments (sets of pixels) where each segment defines an object and a boundary of interest. Segmentation assigns a label to every pixel in an image such that pixels with the same label belong to the same object and share certain common properties. Thus, pixels from the same object (region) are similar to each other with respect to some properties, for instance, similar in color, intensity, or texture. Figure 5.13 shows an example of raw image (left) and its segmentation into objects corresponding to cells (right). In this case, the similarity of white pixels is their low intensity and spatial connectivity forming a large (background) object. The dark pixels in Fig. 5.13 (right) belong to the remaining pixels in the image.

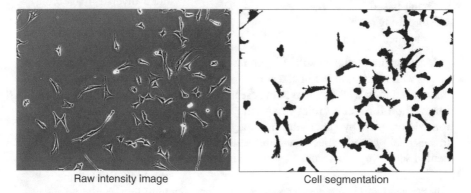

Raw intensity image   Cell segmentation

Fig. 5.13 Example of an National Institutes of Health (NIH) 3 T3 fibroblast intensity image and its corresponding segmentation

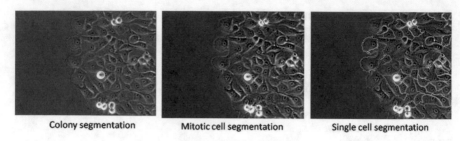

Fig. 5.14 Example of different segmentation of the same image

Segmentation depends on the application and the problem being solved. The same acquired image can be segmented in many ways to solve different problems. Figure 5.14 displays three different segmentation methods applied to the same acquired image to extract three pieces of information: colony boundary (left), boundaries of mitotic cells (middle), or boundaries of each cell (right).

Segmentation classification

The field of image segmentation contains thousands of techniques for multidimensional multivariate images (two- or three-dimensional, gray-scale, color, or hyperspectral variate images). A survey of segmentation methods in [29] classifies them into six major categories:

1. Threshold-based
2. Edge-based
3. Fuzzy theory-based
4. Partial differential equation (PDE)-based
5. Artificial neural network (ANN)-based
6. Region-based

Another survey focused on segmentation methods applied to optical imaging of mammalian cells [30] adds additional categories:

1. Graph-based
2. Morphological
3. Watershed

Image segmentation is still an open research problem, and the number of methods, as well as the number of categories, is a testament to its complexity. An online recommendation system[17] can help researchers choose the appropriate segmentation method for a cell biology experiment using information from past publications. Users start by specifying the imaging modality, objects of interest, image dimensionality, and object measurements and then can choose a segmentation method according to its frequency of use in published papers [30].

---

[17] https://isg.nist.gov/deepzoomweb/resources/survey/index.html

WIPP includes two segmentation techniques:

1. Empirical Gradient Threshold (EGT), a threshold-based segmentation method
2. FogBank, a watershed-based segmentation method

Empirical Gradient Threshold segmentation [31] is a method for separating foreground and background pixels in an image based on empirically established image gradient threshold criteria.

FogBank segmentation [32] is an object separation segmentation method applied to raw images together with their associated binary or labeled mask images. The method has been applied across multiple microscopy imaging modalities and cell lines, and a use case is described in Chap. 3 (migration of breast epithelial cells in sheets).

We introduce threshold-based segmentation category and then describe the EGT methods in detail.

Threshold-based segmentation
Thresholding is a logical operation applied on image intensities to separate foreground from background. There might be some preprocessing and post-processing image surrounding the threshold segmentation. Thresholding can be applied to the image intensities, to the histogram, or to the gradient of the image (edge detection). Thresholding techniques loop over the entire image and set to foreground (value 1) each pixel with intensity meeting the threshold value. For example, if a pixel in a fluorescent image has an intensity value of 500, it will be labeled as foreground if the user is looking at foreground pixels that are higher than 230 in value. A pixel with value of 200 will be set to background (value 0). A binary image is the output after image thresholding where all pixels have a value of 0 or 1 (background or foreground). The key parameter of this method is the determination of the threshold.

Empirical Gradient Threshold segmentation
The method for selecting a threshold value in EGT is presented in three steps:

1. Empirical observations of image gradient values
2. Mathematical model for threshold selection
3. The actual algorithmic steps of EGT

Step 1: Empirical observations of image gradient values
EGT operates on the histogram of the gradient image, and thus it is a histogram shape-based thresholding method as classified by [33]. It derives a model for selecting a threshold value for the gradient of an input image. The model is derived from empirical observations of a reference dataset. The reference dataset consists of 501 validation images with manually determined segmentations and image sizes ranging from 0.36 Megapixels to 850 Megapixels. It includes seven different cell lines and two image modalities (phase contrast and fluorescent).

The gradient of every image in the reference dataset is computed using the Sobel operator [1]. The gradient image is thresholded at every gradient percentile value. The Dice index is computed between every segmented image and the corresponding

Fig. 5.15 Segmentation
comparison of regions A
and B using Dice index

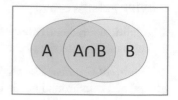

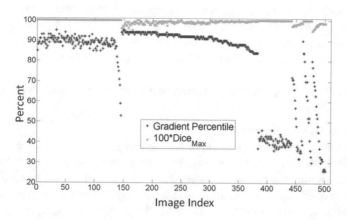

Fig. 5.16 Maximum Dice value (scaled between 0 and 100) and the corresponding gradient percentile threshold for every image in the reference dataset

reference (manual) segmentation. The Dice index is a segmentation accuracy metric
that measures spatial overlap between two segmentations using the following
formula:

$$\text{Dice} = 2 \times A \cap B / (A + B) \tag{5.14}$$

where A and B are the respective areas of the foreground masks as illustrated in
Fig. 5.15. It ranges from 0 (no match) to 1 (perfect match) and is frequently scaled
between 0 and 100.

The best segmentation is chosen as the one with the maximum Dice index per
image dataset. In all reference datasets, the best segmentation was achieved when
the threshold was between the 25th and the 95th percentiles of image gradient values. Figure 5.16 shows a summary of the percentile threshold values across all reference images.

Figure 5.17 shows four examples of normalized gradient intensity histograms
where the 95th, 75th, 55th, and 35th percentiles led to the maximum Dice index,
respectively. In an image where most pixels are background with low-gradient values, higher percentages (e.g., >75%) are needed to reach the correct percentile
threshold for edge detection (Fig. 5.17, 1). In contrast, in an image where most
pixels are foreground, lower percentages (e.g., > 35%) are needed to reach the

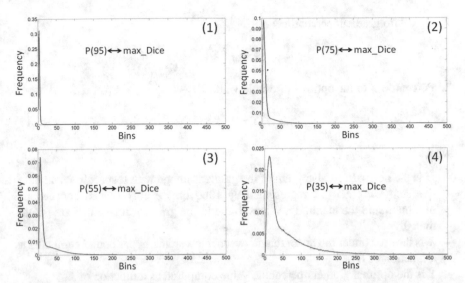

Fig. 5.17 Normalized histogram plots for images where (1) the 95th percentile (2) the 75th percentile, (3) the 55th percentile, and (4) the 35th percentile gave, respectively, the maximum Dice index. The plots are truncated at 500 instead of 1000 on the x-axis to better highlight the difference

correct percentile threshold for edge detection (Fig. 5.17, 3). The difference between the four plots in Fig. 5.17 can be described by how much of the area X under the histogram curve lies to the right of the highest point of the histogram, the mode location.

The background of a biological image usually has low-intensity variations in a small neighborhood surrounding a pixel, which translates to low-gradient magnitudes. Sharp changes in surrounding neighbor intensities around a pixel often correspond to noise in the acquired image. Gradient values for cell or colony edge pixels are usually not the extreme values. Extreme gradient values are more likely from noise. To measure the difference between the four curves in Fig. 5.17, the area X under the histogram curve is computed between a lower bound (*lb*) and an upper bound (*ub*) for each image based on the location of the mode of the histogram, as outlined below.

Step 2: Mathematical Model
The previous section shows that there is a relationship between the histogram distribution and the gradient percentile values from which the threshold is computed. We will model this relationship with three equations relating:

1. Histogram H to area X under the histogram curve between a lower and upper bound

$$X = g(H) \qquad (5.15)$$

2. Area X to gradient percentile Y

$$Y = f(X) \tag{5.16}$$

3. Percentile Y to the optimal threshold value T

$$T = p(Y) \tag{5.17}$$

where

- H is the normalized histogram of the gradient image with respect to its cumulative sum ($sum(H)=1$), represented by 1000 bins evenly spaced between the minimum and the maximum values found in the gradient image that are greater than 0.
- X is the area under the histogram between a lower and upper bound computed as a function of H.
- Y is the optimal gradient percentile value computed as a function of X.
- T is the gradient image intensity threshold value.
- p computes the threshold value T from the percentile value Y. $p(i)$ is the threshold such that $i\%$ of image pixels have intensity gradients less than $p(i)$.

The percentiles are computed from the gradient image without the saturation values (where the gradient is equal to zero). Gradient magnitudes of zero correspond to neighboring pixels in the image where the intensity is the same and thus do not correspond to edge pixels. Lower bounds are always greater than zero. Derivation of the lower and upper bounds is defined in the equations below.

We derived the functions f, g, and p and their respective arguments next.

(a) *Derivation of function $X = g(H)$*

The function g that computes the area X under the histogram curve is modeled as follows:

$$X = g(H) = \sum_{x=lb}^{ub} H(x) \tag{5.18}$$

where lb is a lower bound and ub is an upper bound that will be determined empirically from the mode location. The gradient magnitude mode value generally corresponds to pixels with low-gradient variations (pixels that belong to the background or homogeneous pixels that do not belong to an edge). Since the mode is a statistical value of a histogram, we decided to empirically compute these bounds from an approximated mode location x_{mode}: $lb = n^* x_{mode}$ and $ub = m^* x_{mode}$ with $m > n$. We approximated the mode location x_{mode} using the average of the three highest estimated frequencies. The average mode location value was more accurate than the single maximum peak location and minimized the uncertainty of computing the mode location in the presence of noise and artifacts in the background. The empirical derivation of the lower and upper bounds was made in such a way that it enabled

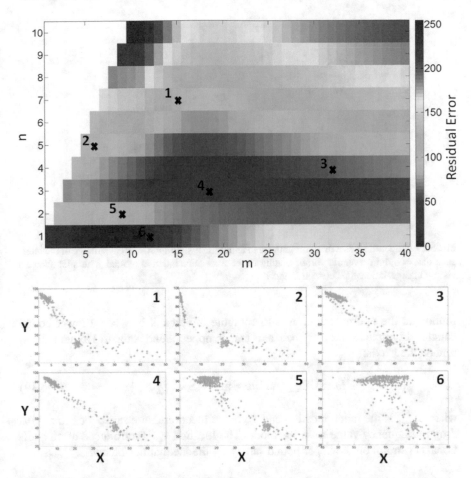

Fig. 5.18 Empirical derivation of the upper and lower bounds of function g. The top portion is the residual error of a linear fit between X and Y. The lower portion displays the plots of the six marked examples, to which the function $Y=f(X)$ will be fit

a known fit (linear if possible) for function f. Therefore, we made an exhaustive search of these bounds looking for linearity of the function f.

Figure 5.18 displays the residual error of a linear fit to the function f color coded between dark blue (lowest error value) and dark red (highest error value). The top portion is the residual error computed with regard to the exhaustive selection of lower and upper bounds as multiples of the mode location. The optimal solution corresponds to the global minimum of the lower and upper bound exhaustive search. The lower portion displays the plots of six marked examples of area X vs. optimal percentile Y, where you can see the linearity. By analyzing the plots in Fig. 5.18, we

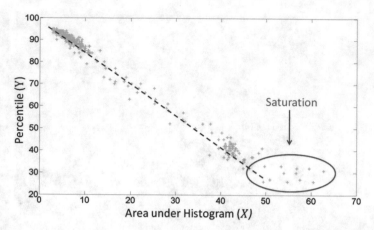

Fig. 5.19 Percentile *Y* as a function of the area under the histogram *X*. For each image of the reference datasets, the percentile corresponding to the max Dice index is plotted. This plot shows a visibly linear relationship between *X* and *Y*

found that the optimal solution is to compute the area X between a lower bound equal to 3× mode location on the x axis and an upper bound equal to 18×mode location on the x axis:

$$lb = 3 \times x_{mode} \text{ and } ub = \max\left(18 \times x_{mode}, x_{cs}\right) \qquad (5.19)$$

where x_{mode} is the mode location and x_{cs} is the location that corresponds to a 95% drop in frequency value from the mode. The location x_{cs} is introduced as an additional constraint on the upper bound satisfying the inequalities below:

$$H\left(x_{cs} + 1\right) > 0.05 \times H\left(x_{mode}\right) \text{ AND } H\left(x_{cs}\right) \le 0.05 \times H\left(x_{mode}\right) \qquad (5.20)$$

(b)Derivation of function Y=f(X)

Figure 5.19 plots the percentile Y corresponding to the maximum Dice index values for all images when computed as a function of the area under the histogram X. This plot reveals a linear relationship between X and Y with a saturation of $Y=25$ for $X \ge 50$. The function f derived empirically from the plot can be written as follows:

$$Y = f(X) = \begin{cases} 95 & X \le s_1 \\ aX + b & s_1 < X < s_2 \\ 25 & s_2 \le X \end{cases} \qquad (5.21)$$

where s_1 and s_2 are derived from the plot with values equal to $s_1 = 3$, $s_2 = 50$.

To compute the linear relationship, we randomly arranged the reference dataset into ten groups of similar size. Nine of the groups are used for training and the remaining one as a validation set. A linear least-squares fit is applied to the training set, and the resulting linear equation is validated on the validation dataset. This process is repeated ten times. The linear function is computed as the average of all ten values and is equal to $a = -1.3517$ and $b = 98.8726$.

(c) *Derivation of Function T=p(Y)*

The image gradient threshold T is derived from the percentile Y by choosing the threshold such that $Y\%$ of image pixels have intensity gradients less than $p(Y)$. Our assumptions for this analysis are that (1) we can segment cells or colonies if edge pixel intensities are different from background intensities and (2) the background is locally uniform. However, there are datasets with images out of focus or with low signal-to noise ratio (SNR). While the relation between X and Y remains linear for these special datasets, the slope of the line drops by a constant factor. A user-defined parameter called "greedy" was introduced because it refers to the "greed" of the background as it claims pixels; this parameter controls the percentile threshold for the entire time sequence of any dataset that falls within these special cases. The greedy parameter is adjusted for only one test image of the entire sequence and is defined as follows:

$$T = p(Y + \text{greedy}) \tag{5.22}$$

with $-50 \leq \text{greedy} \leq 50$, $\text{greedy} \in N$, and $0 \leq Y + \text{greedy} \leq 100$.

The greedy parameter lowers or raises the percentile threshold to capture the missed edge pixels that are in a low- or high-gradient region. Percentiles follow the intensity variations in the image better than just multiplying the current threshold by a factor.

Step 3: Actual algorithmic steps of EGT
The EGT algorithmic steps for segmenting an image are given below:

1. Compute the gradient image G of the raw input image I using Sobel operator.
2. Compute the histogram H of G with 1000 bins.
3. Normalize the histogram with respect to its cumulative sum: $sum(H) = 1$.
4. Average the top three histogram value locations to find an approximate mode location.
5. Compute the area under the histogram X between the lower and upper bounds.
6. Compute $Y = aX + b$.
7. Compute the gradient threshold $T = p(Y)$ and segment the image.
8. Fill holes in the resulting mask that are less than a user-input minimum hole size.
9. Apply morphological erosion with a disk radius of one pixel to clean the noise around the edges.
10. Filter small artifacts that are smaller than a user-specified minimum cell size.

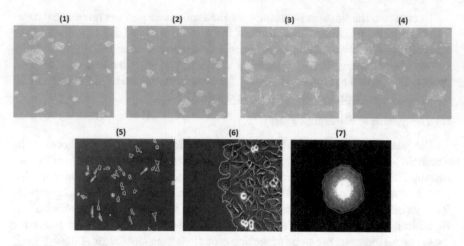

Fig. 5.20 Segmentation results with the contour overlaid on top of the original raw image. Large images (the first four) are zoomed in for better visualization. The cyan color is only for edge highlighting

Figure 5.20 shows examples of segmentation results for the seven reference datasets.

The output of EGT is a binary image. Connectivity analysis needs to be performed on the image to label all connected objects. This is achieved in WIPP by launching "Mask labeling" under "Jobs" menu. After labeling is done, one can compute features and extract interesting data for analysis from a fixed single time point or apply a tracking technique to extract time-series data for analysis. The next section describes tracking the object of interest through time.

5.6.2   Object Tracking Over Time

Object tracking assigns a unique label through time to an object detected by segmentation. Segmentation assigns labels to pixels to form segments at one time point, while tracking associates spatial segments over time. Tracking algorithms make the associations by evaluating similarities between detected objects at adjacent time frames. For example, an object labeled 22 at time t1 and labeled 34 at time t1 + 1 by a segmentation algorithm is the same object, and therefore tracking gives both objects with the same unique global label.

Tracking is useful for extracting dynamic measurements from time-lapse experiments with live cells, such as cell migration, morphology, and lineage development. In general, tracking is performed by assigning a cost between cells from the previous frame and cells from the current frame. The cost value gives a measure of the probability that a cell from the current frame should be tracked to a cell from the previous one. A tracking challenge performed by [34] has a comprehensive list of trackers in the field of biology and particle tracking.

The WIPP tracking algorithm

WIPP uses a tracking algorithm called lineage mapper (LM) to perform tracking across time. LM detects the following 2D dynamic single cell behaviors:

- Migration
- Mitosis
- Cell death
- Cells within sheets
- Cells moving around with high cell-cell contact

While LM was developed for cell biology, it has also been applied successfully for particle tracking. Lineage mapper has five equally important properties:

1. Because it operates on segmented masks, LM does not depend on a specific segmentation method. A user has the choice of any automated or manual segmentation technique; LM takes labeled segmented masks as input. It outputs a cell lineage tree and a set of new labeled masks where each cell is assigned a unique global label.
2. For mitosis tracking, LM utilizes mother cell roundness, mother cell size, daughter size similarity, and daughter aspect ratio measured from segmented images to detect the entire cell cycle across cell divisions.
3. LM uses the overlap information from current and past frames to identify and separate cells mistakenly segmented as a single cell when cell-cell contact occurs.
4. It has fast execution for real-time tracking and manages memory efficiently for large-scale datasets.
5. LM creates fusion lineages by tracking cell or colony merges. It relies on a small number of biologically derived adjustable parameters to achieve high-accuracy tracking.

The core of the LM tracking algorithm performs five main processing steps as shown in Fig. 5.21:

1. Computes a cost function between cells from consecutive frames.
2. Detects cell collisions and separates cells by modifying input images.
3. Performs mitosis event detection.
4. Assigns tracks between cells.
5. Extracts output.

Step 1: Cost function computation

Tracking is performed by assigning a cost (dissimilarity metric) between the i-th cell c_i^t in time frame t and the j-th cell c_j^{t+1} in time frame $(t + 1)$. The cost function $d\left(c_i^t, c_j^{t+1}\right)$ is the sum of three weighted terms computed between cells at consecutive time points t and $t + 1$:

$$d\left(c_i^t, c_j^{t+1}\right) = w_o \times O\left(c_i^t, c_j^{t+1}\right) + w_c \times \delta_c\left(c_i^t, c_j^{t+1}\right) + w_s \times \delta_s\left(c_i^t, c_j^{t+1}\right) \qquad (5.23)$$

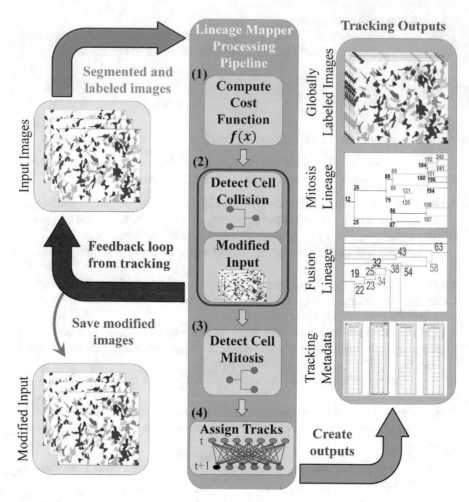

Fig. 5.21 Schematic description of lineage mapper algorithm and output summary data and visualizations

O is the overlap between area of a cell at time t and another at time $t + 1$.

δ_c is the distance metric between cell centroid at time t and another at time $t + 1$.

δ_s is the distance metric between cell size at time t and another at time $t + 1$.

c_i^t is cell i at time t and c_j^{t+1} is cell j at time $t + 1$.

The three terms have multiplicative weights w_o, w_c, and w_s associated with the amount of overlap, the centroid distance, and the size change, respectively. The weights allow the cost function to be tailored for use with different cell lines and image acquisition conditions. It is possible to extend this cost function by adding new terms (like shape metrics, texture-derived metrics, etc.).

Step 2: Collision Detection

Cell collision refers to a group of cells that are correctly detected as individual cells at time t, but are mistakenly segmented as a single cell cluster at time $t + 1$.

Even for very accurate segmentation techniques, spatially adjacent groups of cells can still be mistakenly considered as a single object.

To correct the segmentation mistakes, temporal information about motion of cells can be used to report correct cell tracks. A feedback loop is implemented to separate incorrectly grouped cells into multiple single cell segments by sending the tracking information back to the labeled segments. This option can be disabled by the user to allow object merging or fusion, for example, when tracking cell colonies. When the user enables cell fusion, LM builds a fusion tree where multiple tree branches merge together to form one single branch.

Step 3: Mitosis Detection
The tracking algorithm incorporates four biological indicators that describe mitotic events across most cell lines:

- Mother cells divide normally into two daughter cells during mitosis.
- Before the cell division takes place, the mother cell shape becomes more circular.
- During mitosis, the mother cell does not migrate too much which results in a significant area overlap with its both daughter cells.
- Daughter cells have similar size and shape to each other.

The algorithm models and implements these biological indicators and includes a few user-adjustable parameters to customize each model parameter importance for a given cell line.

Step 4: Track Assignment
After handling mitosis and cell collision, tracks are assigned such that a cell A at time t can share a track with only one cell B at time $t + 1$ and vice versa. The unassigned cells at time t are considered as either leaving the field of view (FOV) or mitotic mothers. The unassigned cells at time $t + 1$ are considered as either entering the FOV or originating from mitosis. The optimal solution to this assignment problem is achieved using the Hungarian algorithm [35, 36]. The Hungarian optimization algorithm minimizes the tracking cost function over all possible tracking assignments after handling mitosis and collision.

After finding the cell tracks for all consecutive pairs of time frames, the frame-to-frame results are combined to produce a complete set of cell life cycle tracks in the time-lapse sequence. Spatially unique cell labels assigned by a segmentation algorithm at each time frame are replaced by spatially and temporally unique track labels that identify each cell at any time in the entire image sequence. The resulting labeled mask is saved in TIFF format. Furthermore, each tracked cell is associated with a confidence index that reflects the magnitude of the cost function during the tracked cell cycle.

Step 5: Tracking Output
The algorithm generates several tracking outputs including the:

- Globally labeled masks where each cell or colony is assigned a unique label for the entire time sequence

- Cell lineage that shows the cell birth, the cell death, the mother-daughter relations, and the number of generations in an image set
- Division and/or fusion lineage that shows the relation between cells that divided and/or collided
- Confidence index

All other outputs can be derived from these primitive ones. For instance, from the globally labeled masks, one can compute the location of each cell centroid and plot the respective migration rate of each cell. All geometric features, for instance, circularity or aspect ratio, can be derived directly from these masks. Similarly, all intensity features can be derived from the masks and the raw images as discussed in the next section.

5.6.3   Image and Object Feature Extractions

After running the segmentation job or the tracking job, one can extract meaningful measurements from each image. These measurements depend on the dimensionality of input images and the semantic meaning of each dimension. We provide classification of image-derived measurements (or feature extractions) based on the type of (1) measurement and (2) spatial region of interest.

Measurement types of feature extractions
In general, measurements (or image-derived features) can be classified into four types:

1. *Intensity (or spectral) features* are extracted from raw intensity images and typically include central moments (average, standard deviation, skewness, kurtosis).
2. *Shape features* are computed from labeled images and focus on capturing spatial measurements, such as size, perimeter, minor and major axis, centroid, circularity, eccentricity, and so on.
3. *Textural features* are calculated by transforming intensities inside of regions of interest into a space where one can make direct measurements of spatially repeating patterns. Among those transformations, the most common ones are Gabor, Fourier, law's, and gray-level co-occurrence matrix (GLCM) transforms. Examples of GLCM-derived textural features are contrast, homogeneity, energy and entropy.
4. *Motion features* are temporal measurements of object tracks extracted from time-lapse image sequences and include cell migration rate, cell growth rate, or temporal change of protein expression.

Spatial types of feature extractions
Feature extraction is accomplished by algorithms that ingest raw intensity images and/or labeled images and/or multiple labeled images. Example inputs to a feature extraction algorithm are shown in Fig. 5.22. In general, there are three types of spatial measurements:

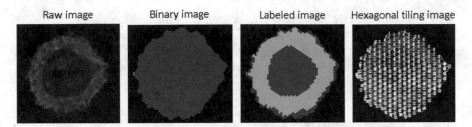

Fig. 5.22 Left to right: raw intensity image; binary image separating cell foreground and background using edge detection; labeled image containing labels for cell subdivided into nucleus, cytoskeleton, and focal adhesion subregions; and hexagonal tiling image obtained by spatially tiling the cell foreground region using hexagonal tessellation

1. *Image measurements* where it is assumed that one FOV is a sample of an object of interest.
2. *Spatially global object measurements* where an object of interest is defined by a foreground mask (Fig. 5.22, edge detection). Multiple objects are described by a labeled mask (Fig. 5.22, segmented image). For example, objects can be cell colonies, cells, or cell layers.
3. *Spatially local object measurements* where an object has spatially varying properties and therefore an object mask and its tessellation mask are needed to compute local homogeneity measurements (Fig. 5.22 hexagonal tiles within a cell).

In WIPP, the feature extractor types are supported by integrating several widely used software packages. The current implementation includes primarily intensity, shape, and textural features. The spatial feature extractors leverage segmentation results (EGT and FogBank) and tessellation results. The tessellation job is designed to partition a user-defined image area into hexagonal or rectangular segments with unique labels. The combination of a labeled mask from segmentation (e.g., cell colony segments) and a labeled mask from tessellation (e.g., hexagonal segments) can be used as input to the web feature extraction module in order to compute features with respect to hexagonal segments inside of a cell colony. This can be used to evaluate cell colony homogeneity.

5.7   Image Intensity Scaling and Pyramid Building Algorithms

Current web browser technologies do not support image rendering when a single channel has more than 8 bits per pixels (BPP). This limitation must be overcome by scaling the pixel intensities, because most images acquired by optical microscopes are in 12 or 16 BPP format. Scaling the raw images before launching pyramid building is recommended since the pyramids are used directly for visualization in a web browser. Without the intensity rescaling, the visualization might not render correctly in the browser (i.e., images might be too dark). All other computations should be performed on the full dynamic range images. We will next cover the options for the intensity scaling and pyramid building algorithms.

5.7.1   Image Intensity Scaling

The goal of image intensity scaling is to compress larger image bit depths to 8 BPP while preserving salient image features for visual inspection. From a mathematical viewpoint, the challenge lies in defining a model for mapping the original intensity values into a smaller number of bins while preserving the magnitude order of intensities. From an application viewpoint, the challenge lies in choosing a model that allows a user to interactively define transformation parameters so that the user can see salient features of interest. This perspective also includes some human factors since some scientists prefer linear scaling to preserve linear relationships between intensities.

In photography, intensity scaling is also called tone mapping. It can be global or local to address illumination changes (e.g., objects in shadows and in sunlight). The global mapping applies a scaling model (i.e., tonal curve) to all pixel intensities. The local mapping is different depending on the pixel location. In microscopy, the illumination changes are not significant, and the main concerns are human perception and display limitations. To address these concerns, intensity scaling algorithms implement either linear or nonlinear global mapping. The choice of a model for intensity mapping is typically supported by human subject studies.

WIPP contains two implementations of intensity scaling to 8 BPP: (1) truncation and (2) gamma correction. Figure 5.23 displays an example for each of the two intensity scaling methods.

Truncation

This method computes an intensity histogram and assigns (a) zeros to all pixel intensities that are below the threshold defining 1 % of histogram and (b) 255 to all pixels above the threshold defining 99 % of histogram. This method maps the rest of the intensities linearly between these two thresholds.

Figure 5.24 (left) shows the intensity histogram of the raw image where the value ranges from 216 to 3318, and after truncation the intensity values now ranges from 0 to 65 535 with a saturation of 1 % of pixels around the 1 % and the 99 % of the intensity values. Note that the new pixel intensities occupy the full 16 BPP range, not the 8 BPP range. This is because the pyramid building job rescales all input images to 8 BPP using a linear rescaling of the full input image dynamic range. Therefore, for correct visualization the intensity scaling job needs to stretch images to use their full dynamic range.

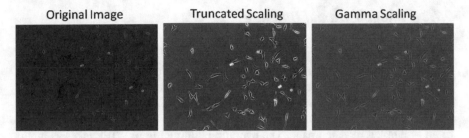

Fig. 5.23 Scaling method example highlighting scaling effect on the original image

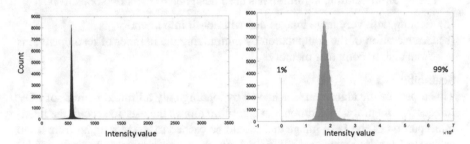

Fig. 5.24 Histogram of raw image intensity (left) and the truncated one (right)

Gamma correction

This nonlinear method applies the following transformation to each pixel intensity:

$$I_c = AI^\gamma \tag{5.24}$$

where I_c is the scaled images, I is the original image, γ is the gamma correction factor ($\gamma = 0.1$ in WIPP), and A is a constant (usually equal to 1). The resulting image is brighter or darker than the original image depending on whether the gamma value is smaller or larger than 1.

If image scaling job was not executed before launching a pyramid job, images with more than 8 BPP will likely appear dark in the web browser. To overcome the problem, one can view the dark pyramids in a Deep Zoom viewer by clicking on the button "Modify Filters" and then choose "Exposure" or "Brightness" or "Contrast" filters. These filters allow a user to interactively find the right settings for visual inspection. The disadvantage of the browser-based filter adjustment is that the computation may slow down older computers.

5.7.2 *Image Pyramid Building*

Memory constraints on pyramid building

Chapter 4 ("Representation of Large Images") introduced the concept of image pyramid representations for big images. Section "Image Mosaicking" in this chapter mentioned the RAM limitations when big images are reconstructed in memory. There is a need for the pyramid building algorithm to start with raw FOVs instead of an assembled mosaic image. The pyramid building job in WIPP takes as inputs both a collection of FOV images and the associated vector file containing FOV position information in the coordinate system of the mosaic image. The stitching job must be executed before pyramid building to obtain the position vector file.

The pyramid building algorithm has been designed to meet two objectives:

(a) Scaling with very large images that will not fit into memory
(b) Acceleration of the computation by minimizing the number of read operations from disk to computer memory

Scalability

This aspect of the algorithm is achieved by loading only a limited number of FOV images into memory. In addition, the algorithm allows the user to specify the maximum portion of the FOV image that should be cached in RAM ranging from 0 (no cache) to 1 (entire image cached). By default, the output pyramid tile size is (254 × 254) pixels with one pixel overlap.

Number of read operations

Figure 5.25 shows color-coded tiles in each layer of a pyramid representation where the same color indicates related tiles between layers p and (p-1). Figure 5.26 illustrates the numerical order of building pyramid tiles to minimize the number of read operations. Each image region forming a tile at the maximum zoom level p is read only once. A tile at level p-1, which is composed of four tiles at level p, is computed as soon as enough tiles at level p have been computed.

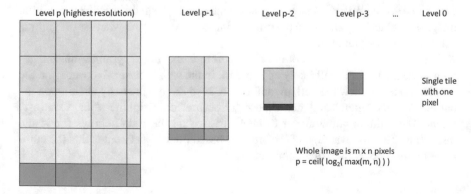

Fig. 5.25 Pyramid representation

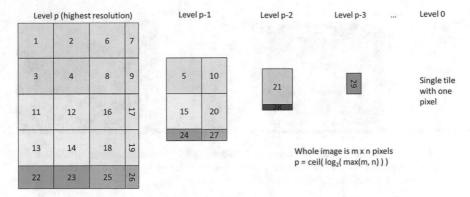

Fig. 5.26 The order of building pyramid tiles to minimize the number of read operations

Pyramid building – pseudo-code

An implementation of algorithm is available from the Git repository[18]; its pseudo-code is provided below:

```
Tile buildTile(int level, int row, int column)
    if (level == p)
        read tile area from raw FOVs using stitching vector and
        perform blending
        save it to disk
        return tile
    else
        Tile topLeft = buildTile(level + 1, row * 2, column * 2);
        Tile topRight = buildTile(level + 1, row * 2, column * 2 + 1);
        Tile bottomLeft = buildTile(level + 1, row * 2 + 1, column * 2);
        Tile bottomRight = buildTile(level + 1, row * 2 + 1, column * 2 + 1);
        Assemble the 4 tiles in one image and scale it down to create
        the tile
        save it to disk
        return tile

void buildPyramid()
    buildTile(0, 0, 0);
```

This recursive algorithm also supports parallel execution of pyramid tile building. A pyramid can be generated using a network file system, but the file transfers significantly slow down the computation. There are also several options related to blending spatially overlapping areas of adjacent FOVs, for instance, averaging or

---

[18] https://github.com/usnistgov/pyramidio

Fig. 5.27 A spatial view of a historical artifact (https://isg.nist.gov/deepzoomweb/data/materiale-nergylevels) imaged using an electron microscope (left) and its re-projected spectral view $[y, \lambda]$ for one of the x-cross sections (right)

using extreme intensities at overlapping locations. The optimal solution is to use a look-up table for the FOVs covering the overlapping area, read the files, and blend them using one of the options before pyramid tiles are computed.

5.7.3   Re-projection of a Pyramid Set

Typically, large image collections form a 3D volume with spatial $[x, y, z]$, temporal $[t]$, and/or spectral $[\lambda]$ dimensions. Scientists need to inspect the acquired TB-sized images in the context of their spatial and spectral surroundings. For example, cell touching each other might behave differently than cells that are spatially far apart. Similarly, a spectral composition of materials varies spatially and must be inspected across spatial cross sections to understand material heterogeneity as illustrated in Fig. 5.27. The neighboring contextual image information must be viewed in three orthogonal planes: $[x, y]$, $[x, z/t/\lambda]$, and $[y, z/t/\lambda]$. To enable scientists to inspect and measure TB-sized 3D volumes, there is a need for fast orthogonal re-projections to facilitate interactive view changes.

The 3D image volume can be represented as a set of image frames or a set of image pyramids. For both representations, re-projection algorithms are needed as illustrated in Fig. 5.27. The re-projection algorithms reshuffle pixels as illustrated in Fig. 5.28 for a set of 2D image frames and in Fig. 5.29 for a set of image pyramids.

We provide the top-level algorithmic workflow steps below. Without any loss of generality, we analyze x-y-z volumes (row-column-depth) represented as a set of x-y frames that are re-projected to a set of x-z frames.

Pseudo-code: Image frame set to image frame set re-projection
- Load 2D images into RAM.
- FOR all columns in the input 2D frames

  - Copy the same index columns from each 2D frame and put them into a new 2D image according to the input order of 2D frames.
  - Write the new 2D image to disk.

- END

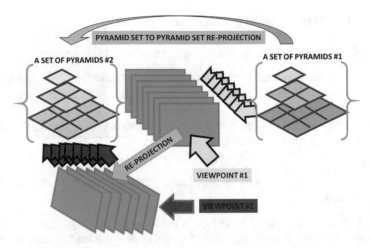

Fig. 5.28 Re-projection of large 3D images to achieve two orthogonal viewpoints

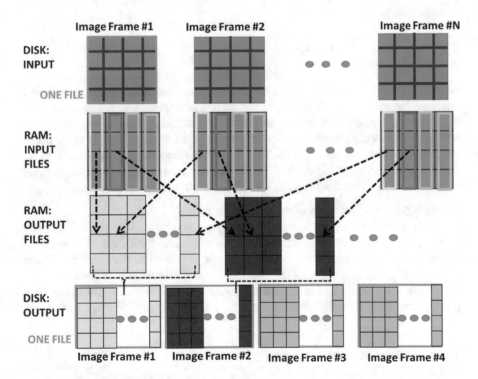

Fig. 5.29 Re-projection from a set of 2D image frames

Pseudo-code: Pyramid set to pyramid set re-projection
* Load 2D image tiles at the zoom level one into RAM.
* FOR all columns and image tiles in the input pyramid sets

    - Copy the same index columns from the same index image tile from each pyramid set and put them into a new 2D image tile according to the input order of pyramid sets.
    - Write the new 2D image tile to disk inside of the output pyramid set numbered according to the column index.

* END
* Complete building the full pyramid sets from the zoom level one tiles by down-sampling.

Based on the experimental results [37], the re-projection algorithm for pyramid sets is more efficient than the algorithm for sets of image frames when executed on distributed computational resources. The main advantage of the pyramid-based re-projection is that it does not require as much computer memory and immediately generates the pyramid representation for viewing.

5.8   Supervised Algorithms

All algorithms presented in this chapter are based on some underlying model designed by a developer. There is an option of integrating data-driven supervised algorithms where the models are automatically learned from the presented annotated data. These algorithms have been shown to be accurate, especially those based on deep learning models [38, 39]. They have been successfully applied to many image segmentation problems. We provide a brief comparison of advantages and challenges of deep learning models for segmentation tasks next.

Advantages and challenges of deep learning models
Complex deep learning models can be designed and trained by providing many training examples. Table 5.7 summarizes advantages and challenges of deep learning models. On one hand, the models represent a transition from the past linear model to highly nonlinear models, introduce automatic feature engineering, and parallelize the training and inference processes well on Graphics Processing Unit (GPU) hardware. On the other hand, the models do not provide visual nor

Table 5.7 Advantages and challenges of deep learning models

Deep learning advantages	Deep learning challenges
Linear → nonlinear models	Difficult to inspect models visually
Automatic feature engineering	Lack of theoretical understanding
Computations parallelize well on GPU hardware	Hand-designed architectures
	Training needs lots of data

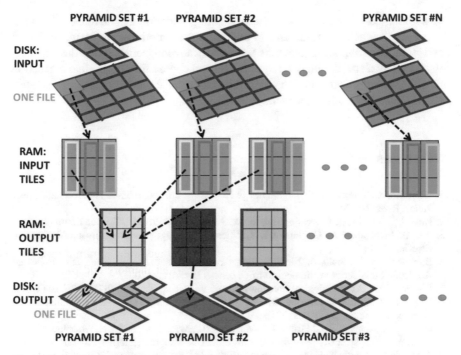

Fig. 5.30 Re-projection from a set of pyramids

mathematical insights about the input-to-output relationship, and the model architectures must be hand-designed. One should be aware that the models work for stationary phenomena (time-invariant models) and for a particular spatial resolution of the training dataset. It is also imperative to choose balanced training data (i.e., equal representation of all classes) since the models reinforce majority.

5.9 Summary

The image processing algorithms documented in this chapter have been primarily designed for analyzing microscopy images in cell biology and tissue/cell pathology (histo- and cytopathology). One can learn more about bioimage processing by:

1. Reading widely used textbooks
2. Installing software implementations used for bioimage analyses
3. Loading or converting image files to an OME standard file format
4. Applying image processing operations to acquired images

Image processing algorithms can be adapted to a variety of microscopy image contents. We introduced algorithms used in common analyses of microscopy images including image correction, stitching, segmentation, tracking, feature extraction, and pyramid-based visualization. Before adapting an algorithm, one must understand the purpose of each algorithm and its underlying assumptions. For instance, image correction algorithms assume that a user is knowledgeable about the artifacts introduced by a specific imaging system and can choose the appropriate correction models. Similarly, segmentation algorithms incorporate models of objects of interest that depend on each experiment.

References

1. Gonzalez, R.C., Woods, R.E.: Digital Image Processing, 3rd edn. Pearson Prentice Hall. Upper Saddle River, NJ (2007)
2. Russ, J.C.: The Image Processing Handbook, 3rd edn. CRC Press LLC, Boca Raton (2002)
3. Wu, Q., Merchant, F.A., Castleman, K.R.: Microscope image processing. Boston/Burlington, Elsevier (2008)
4. Bankman, I.: Handbook of Medical Image Processing and Analysis, 2nd edn. Eslsevier, Academic Press Series in Biomedical Engineering, Burlington (2008)
5. Eliceiri, K.W., et al.: Biological imaging software tools. Nat. Methods. 9(7), 697–710 (2012)
6. Miura K. Bioimage Data Analyses. Miura K, editor. Viley-VCH, Verlag-GmbH. 69469 Weinheim, Germany: Olympus; 2016
7. Bajcsy, P., Kooper, R., Marini, L., Minsker, B., Myers, J.: A Meta-Workflow Cyber-infrastructure System Designed for Environmental Observatories [Internet]. Urbana, IL; 2005. Available from: http://isda.ncsa.uiuc.edu/peter/publications/techreports/2005/meta-workflow-approaches.pdf
8. Bankhead P. Analyzing fluorescence microscopy images with ImageJ [Internet]. Heidelberg University, Germany; 2014. Available from: http://go.qub.ac.uk/imagej-intro
9. Chalfoun, J., Majurski, M., Bhadriraju, K., Lund, S., Bajcsy, P., Brady, M.: Background intensity correction for terabyte-sized time-lapse images. J. Microsc. 257(3), 226–238 (2015)
10. Sharma, A., Singh, J.: Image denoising using spatial domain filters: a quantitative study. 2013 6th Int. Congr. Image Signal Process. 1(Cisp), 293–298 (2013)
11. Wyawahare, M.V., Patil, P.M., Abhyankar, H.K.: Image registration Techniques : an overview. Pattern Recogn. 2(3), 11–28 (2009)
12. Zitova, B.: Image registration methods: a survey. Image Vis. Comput. 21(11), 977–1000 (2003)
13. Brown, M., Lowe, D.G.: Automatic panoramic image stitching using invariant features. Int. J. Comput. Vis. 74(1), 59–73 (2006)
14. Bajcsy, P., Lee, S.-C., Lin, A., Folberg, R.: Three-dimensional volume reconstruction of extra-cellular matrix proteins in uveal melanoma from fluorescent confocal laser scanning microscope images. J. Microsc. 221(Pt 1), 30–45 (2006)
15. Can, A., Stewart, C.V., Roysam, B., Tanenbaum, H.L.: A feature-based technique for joint, linear estimation of high-order image-to-mosaic transformations: Mosaicing the curved human retina. IEEE Trans. Pattern Anal. Mach. Intell. 24(3), 412–419 (2002)
16. Chow, S., et al.: Automated microscopy system for mosaic acquisition and processing. J. Microsc. 222, 76–84 (2006)
17. Saalfeld, S., Cardona, A., Hartenstein, V., Tomancák, P.: As-rigid-as-possible mosaicking and serial section registration of large ssTEM datasets. Bioinformatics. 26(12), 57–63 (2010)
18. Tsai, C.-L., et al.: Robust, globally consistent and fully automatic multi-image registration and montage synthesis for 3-D multi-channel images. J. Microsc. 243(2), 154–171 (2011)
19. Preibisch, S., Saalfeld, S., Tomancak, P.: Globally optimal stitching of tiled 3D microscopic image acquisitions. Bioinformatics. 25(11), 1463–1465 (2009)

20. Argyriou, V.: A study of sub-pixel motion estimation using phase correlation. Br. Mach. Vis. Assoc. 17th BMVC. 1–10 (2006)
21. Bican, J., Flusser, J.: 3D rigid registration by cylindrical phase correlation method. Pattern Recogn. Lett. **30**(10), 914–921 (2009)
22. Davis, J.: Mosaics of scenes with moving objects. In Proceedings of 1998 IEEE Computer Society Conference on Computer Vision and Pattern Recognition, pp. 354–360 (1998)
23. Emmenlauer, M., et al.: XuvTools: free, fast and reliable stitching of large 3D datasets. J. Microsc. **233**(1), 42–60 (2009)
24. Koshevoy, P., et al.: Automatic mosaicking and volume assembly for high-throughput serial-section transmission electron microscopy. J. Neurosci. Methods. **193**(1), 132–144 (2011)
25. Steckhan, D., Bergen, T., Wittenberg, T., Rupp, S.: Efficient large scale image stitching for virtual microscopy. Conf. Proc. IEEE Eng. Med. Biol. Soc. **2008**, 4019–4023 (2008)
26. Chalfoun, J., Majurski, M., Blattner, T., Keyrouz, W., Bajcsy, P., Brady, M.: MIST accurate and scalable microscopy image stitching method with stage Modeling and error minimization. Nat. Sci. Reports. **7**, 1–10 (2017) Available from: https://www.nature.com/articles/s41598-017-04567-y.pdf
27. Blattner, T., Keyrouz, W., Chalfoun, J., Stivalet, B., Brady, M., Shujia, Z.: A hybrid CPU-GPU system for stitching large scale optical microscopy images. In Parallel Processing (ICPP), 2014 43rd International Conference on, 2014, pp. 1–9
28. Gonzales, R., Woods, R.E.: Digital Image Processing, 3rd edn. Pearson (2007)
29. Khan, W.: Image segmentation techniques: a survey. J. Image Graph. **1**(4), 166–170 (2014)
30. Bajcsy, P., et al.: Survey statistics of automated segmentations applied to optical imaging of mammalian cells. BMC Bioinformatics. **16**(330), 1–28 (2015)
31. Chalfoun, J., Majurski, M., Peskin, A., Breen, C., Bajcsy, P., Brady, M.: Empirical gradient threshold technique for automated segmentation across image modalities and cell lines. J. Microsc. **260**(1), 86–99 (2015)
32. Chalfoun, J., Majurski, M., Dima, A., Stuelten, C., Peskin, A.: FogBank : a single cell segmentation across multiple cell lines and image modalities. BMC Bioinformatics. **15**(431), 12 (2014)
33. Sezgin, M., Sankur, B.: Survey over image thresholding techniques and quantitative performance evaluation. J. Electron. Imaging. **13**(1), 146–165 (2004)
34. Chenouard, N., et al.: Objective comparison of particle tracking methods. Nat. Methods. **11**(3), 281–289 (Mar. 2014)
35. Bise, R., Yin, Z.Kanade, T.: Reliable cell tracking by global data association. In Biomedical Imaging: From Nano to Macro, 2011 IEEE International Symposium on, 2011, pp. 1004–1010e
36. Dasgupta, D., Hernandez, G., Garrett, D., Vejandla, P.K., Kaushal, A., Yerneni, R., et al.: A comparison of multiobjective evolutionary algorithms with informed initialization and kuhn-munkres algorithm for the sailor assignment problem. In: Proceedings of 2008 GECCO Conference companion Genetics Evolution Computation – GECCO '08, pp. 2129–2134. ACM Press, Atlanta (2008) Available from: http://portal.acm.org/citation.cfm?doid=1388969.1389035
37. Vandecreme, A., Bajcsy, P., Ritchie, N.W.M., Scott, J.H.J.: Interactive analysis of terabyte-sized SEM-EDS Hyperspectral images. Microsc. Microanal. **20**(Suppl 3), 654–655 (2014)
38. Garcia-Garcia, A., Orts-Escolano, S., Oprea, S., Villena-Martinez, V., Garcia-Rodriguez, J.: A review on deep learning techniques applied to semantic segmentation. Arxiv. **arXiv**(1704), 1–23 (2017) Available from: http://arxiv.org/abs/1704.06857
39. Zhao, B., Feng, J., Wu, X., Yan, S.: A survey on deep learning-based fine-grained object classification and semantic segmentation. Int. J. Autom. Comput. **14**(2), 119–135 (2017)

Chapter 6
Interoperability Between Software and Hardware

6.1 Hardware Options for Accelerating Computations

Introduction to Hardware and Software Interoperability

Big image data can require significant processing time. Software and hardware can be exploited to shorten this time. In addition to increasing microscope acquisition speed and exponential data growth, information technology is also rapidly advancing. To leverage this new information technology, one must deal with a diversity of hardware and software interfaces. Software must be written with hardware specifications in mind and must integrate with other existing software. These software engineering activities are driven by our goals not only to shorten computation time but also to minimize the effort of building big data analytic solutions. While WIPP can be deployed on a variety of hardware, its execution performance will depend on the selection of that underlying hardware. We present next the hardware options with their pros and cons.

Advanced hardware options

In general, advanced hardware can accelerate big data processing by minimizing the time for read/write operations, network data movement, and data processing. Based on hardware specifications, minimization of the overall execution time can be achieved by:

(a) Keeping all data in RAM during computations instead of transferring them from hard drives
(b) Utilizing processors with higher clock speeds to increase the number of computations per time unit
(c) Using high bandwidth communication buses to decrease data transfer time
(d) Parallelizing read/write, data movement, and computation operations
(e) Utilizing special computational hardware that is more efficient than available central processing units (CPUs)
(f) Using nonvolatile memory (NVM) express devices with solid-state drives (SSDs) for faster data access

© Springer International Publishing AG 2018
P. Bajcsy et al., *Web Microanalysis of Big Image Data*,
https://doi.org/10.1007/978-3-319-63360-2_6

To implement big data performance acceleration, there are currently four hardware options for principal investigators or small research teams:

1. Access a supercomputer with a very large RAM, fast connectivity, and powerful computational nodes (CPUs and GPUs).
2. Buy a high-end desktop computer with advanced hardware (large RAM, powerful processors, fast buses), and fully utilize this hardware for faster computation.
3. Access multiple computers (either in a cluster or cloud) and exploit them in parallel.
4. Utilize additional computer hardware (e.g., GPUs, FPGAs, or NVM SSDs) designed specifically for efficient computation and fast storage, and then implement custom software for this hardware.

All of the above options require integrating software and hardware by redesigning existing software or designing new software to fully utilize the hardware. We will briefly describe each of the four options.

Access a supercomputer
The supercomputers are available to a small percentage of researchers because of their limited availability. The key difference between the big data and supercomputing communities is the ratio of data size to the number of computations per data point. The primary focus of the supercomputing community is on computations whose results require little input data but a very large amount of computation. Numerical simulations are an example. The big data community, on the other hand, is interested in information extraction and data-driven modeling from very large datasets. Information extraction might not require a large number of computations per data point, but the number of data points is very large. Unlike information extraction, data-driven modeling can also demand a large number of computations per data point. The ratio of data size to the number of computations for big data computations leads us to emphasize data distribution in order to parallelize computations.

Buy a high-end desktop computer
One can invest in a personal computer with advanced hardware (faster processor, more RAM, faster disk access, etc.). This approach is expensive and is limited by the best available hardware. The currently available multicore processors require developing algorithmic implementations that can leverage them.

We illustrate this approach in Fig. 6.1 by drawing a parallel between the throughput of big data processing and the throughput of cars on automobile highways. The colored cars are types of computations, and a highway lane is one processor with its bus connecting the processor to all needed data. Buying a multicore computer to run single-threaded code can be compared to paying for a multiple lane highway and then using only one lane. Our throughput metric is the number of computations per time (or number of cars per time interval). It is apparent that the software affects performance by defining how the big data input is handled by hardware during computation.

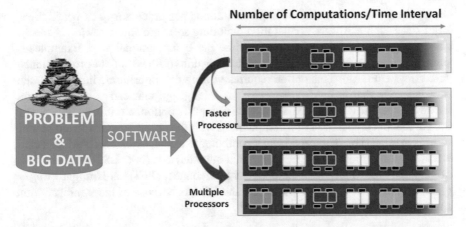

Fig. 6.1 Accelerating big data problem computations by purchasing a faster processor or by redesigning software and buying a computer with multiple processors

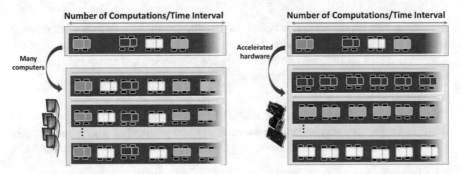

Fig. 6.2 Accelerating big data problem computations by redesigning software to run on multiple computers in parallel (left) or on accelerated hardware and attaching the hardware to a computer

Access multiple computers

Another option for accelerating big data computations is to use multiple computers in parallel as illustrated in Fig. 6.2 (left). This option requires designing algorithmic implementations together with the software that would move parts of the data and parts of the computations to multiple computers and then collect the results on a single computer. Critical components in this case are (a) the decomposition of a computation into steps that can be executed in parallel and (b) the partition of input data into chunks that are needed by each step. We will focus primarily on the Hadoop framework [1] for such algorithmic implementations.

Utilize additional computer hardware

The last option is to use additional specialize hardware for accelerating computations as illustrated in Fig. 6.2 (right). Special computer hardware can perform some computational steps more efficiently than a general-purpose CPU. Figure 6.2 (right)

shows color-coded grouping of implementations per processor (cars per highway lane) since each hardware acceleration including software and hardware is specifically designed to minimize execution time per computational step. Examples of such computer hardware are graphics processing units (GPUs) or field-programmable gate arrays (FPGAs). This option requires writing the algorithmic implementation in a language that is hardware-specific or a language that can be compiled into hardware-specific code. Like the multiple computer option, a critical component is the parallel decomposition of a computation with the accelerated hardware specifications in mind (i.e., bus speed from a computer to a card, available RAM on a card, etc.). The use of nonvolatile memory solid-state devices (NVM SSDs) in the form of standard-sized Peripheral Component Interconnect (PCI) cards might bring as much overall time savings as the GPUs and FPGAs because of the significant time spent reading and writing big image data.

6.2   Implications of Big Data Attributes

In any big data experiment, it is beneficial to estimate the rate of image acquisition and the size of the image collections to be processed within a given time interval. These estimates allow a user to plan hardware purchases and select big data solutions according to attributes of big data.

Attributes of big data

The distribution of big data across multicore processors, multiple computers, and multiple cards with accelerated hardware should be performed with respect to big data attributes. The attributes defining big data are described as four Vs and include:

- Volume
- Velocity
- Variety
- Veracity

Volume is the data size on disk measured in bytes. Velocity is the speed in which data can be accessed, for example, the data acquired by a microscope scanning a slide. Variety refers to structured or unstructured organizations of different types of data (e.g., structured XML key-value pairs or unstructured text from blogs). Veracity reflects whether the data are accurate or not (calibrated vs. uncalibrated microscope images, inaccurate metadata about images).

Some practitioners[1] add additional attributes such as:

- Variability
- Visualization
- Value

---

[1] https://www.impactradius.com/blog/7-vs-big-data/

These extra attributes are less frequently cited. Variability refers to inconsistent meaning or labeling of data, for instance, two microscopes reporting dimensions of a pixel as "pixel size" or "pixel dimension" (important for data integration). Visualization refers to conveying the meaning in a pictorial or graphical format rather than in a spreadsheet numerical format. Value is the ultimate measure of the information gain after addressing volume, velocity, variety, variability, veracity, and visualization.

Big data solutions

The application-specific big data attributes have implications on storage, network, and computational hardware specifications which lead to solutions that operate from the scale of an imaging lab to the scale of a large IT company. For example, by considering the volume and velocity attributes, a user must plan for big data storage that (a) can handle the current size and the data growth (scale) and (b) can provide a fast access to the data (the input/output operations per second (IOPS)). We will describe the implications of big data attributes on hardware and software at commercial and imaging laboratory scales and the role of interoperability and standards for big data solutions.

Big data solutions at commercial scale

In a commercial space, large IT companies create *hyperscale computing environments*, where the term "hyper" refers to excessive and "scale" refers to data growth.[2] The hyperscale computing environment is achieved by assembling commodity servers for petabytes of data with direct-attached storage (DAS) and storage redundancy at the level of the entire computer/storage unit. In order to serve millions of users with thousands of applications, these environments are reducing storage latency by having Peripheral Component Interconnect Express (PCI-E) flash storage and running analytic engines like Hadoop, NoSQL, and Cassandra.

Big data solutions at imaging laboratory scale

For researchers running big microscopy image experiments, the volume attribute of big data is typically at the scale of terabytes, and the velocity is between 1 MB/s and 100 MB/s (see Chap. 1, Fig. 1.6). This implies that network-attached storage (NAS) and clustered NAS with multiple terabyte-sized storage capacity might be sufficient and cost-efficient. As software for clustering NAS improves to deliver petabyte/parallel file system capability, this solution may meet the requirements for *scaling up* (adding more disks to the same disk controllers). With the growing number of cloud providers, there is also an option of *scaling out* by creating a distributed storage (requesting more nodes in the computer cloud to provide a larger aggregated storage). Finally, the need for collocating data with computational resources suggests an option of bringing data storage and computers into one geographical locations with fast network connectivity, cooling, and sufficient power. This option is provided by commercial vendors and is referred to as *colocation centers*.

---

[2] http://searchstorage.techtarget.com/podcast/Understanding-stripped-down-hyperscale-storage-for-big-data-use-cases

Interoperability and Standards
The attributes of big data have been one of the main topics discussed in the NIST
Big Data Public Working Group (NBD-PWG).[3] This working group has focused on
Big Data Definition, Big Data Taxonomies, Big Data Use Cases and Requirements,
Big Data Security and Privacy, Big Data Reference Architecture, Big Data Standards
Roadmap, Big Data Reference Architecture Interface, and Big Data Adoption and
Modernization. The work of NBD-PWG has led to reports that address the interop-
erability of big data solutions ever since "the world was awash with over 800 exa-
bytes of data and growing" as estimated in 2010 by Thomson Reuters.[4]

6.3   Execution Times of Computation Over Big Image Data

Response time-based classification of computations
Based on the storage considerations for the image data volume and velocity attributes, one must decide on the hardware and software. Processing is classified as
either *off-line* or *interactive*. In other words, interactive computations are expected
to take the time of a mouse click (i.e., seconds), while off-line computations can
take minutes, hours, and days. This classification is focused on a response time to
user inputs. For systems designed for interactive, discovery-type, analyses, and
image explorations, it is important to:

1. Classify computations based on their execution time requirements.
2. Decide on strategies to meet the time requirements.
3. Support the chosen strategy by evaluations of computational complexity.
4. Collect measurements of actual execution times to validate the system
   performance.

In the next two subsections, we focus on (1) meeting execution time requirements
and (2) estimating and measuring execution time over big image data.

6.3.1   Meeting Execution Time Requirements

Local vs. distributed computing
To meet performance requirements, analyses can be executed on a local computer
or on a set of distributed computational nodes. For local analyses, data are transferred to a local computer, and computations are constrained to a local desktop. For
distributed analyses, data chunks are sprayed across multiple computer cluster or
cloud nodes. Computations are executed by processing data chunks followed by
merging partial results into an overall result. If advanced hardware is available on
local or distributed computational resources, then the data must be transferred to it

---

[3] https://bigdatawg.nist.gov/

[4] http://archive.annual-report.thomsonreuters.com/2010/

(e.g., to GPU or FPGA memory). The key aspect of these processing options is to co-localize big data with the most processing power. The execution time depends on the time to move data, the availability of processing resources, and the software design for fully utilizing each hardware device.

On-demand computing
Another aspect of the performance requirements is the number of concurrent users. Users launch computations in an unpredictable pattern on web systems like WIPP. If the computational resources are busy, then computations will wait in a queue and cause long execution times. During these "on-demand" computations, the hardware and software solutions must have access to additional computational resources to meet performance requirements. The resources could be plug-and-play hardware devices or new virtual machines (i.e., elasticity of cloud computing).

Client-server computing as a type of on-demand computing
In the context of client-server web systems, the web server accesses distributed computing resources to execute analyses (see Fig. 6.3). The resource allocation for distributed analyses depends on computational demands. It can be elastic as described above. In comparison to execution using distributed computing resources, local analyses can be executed on a client computer or on a main web server. The challenge lies in co-localizing data and processing power. For example, if images are already transferred to a client for visualization, then the images might immediately be processed on the client to save the time to process and move the data from the web server to the client. If a client does not have sufficient processing power to meet performance requirements, then the images may be processed on a web server and then transferred to a client. The choice must be made during development by making assumptions about the server-client transfer rates and relative comparison of server-client processing power. The current implementation of WIPP assumes the classification of computations as illustrated in Fig. 6.3.

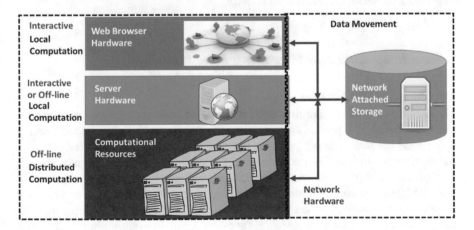

Fig. 6.3 Interactive versus off-line and local versus distributed computations in the context of the WIPP client-server system

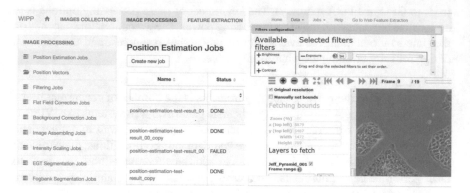

Fig. 6.4 Left – computational jobs executed by a web server on distributed computational resources. Right – image filtering operations executed by a web client on a local client machine

Strategies for off-line vs interactive computations in WIPP

In WIPP, all computations accessible from the Image Processing or Feature Extraction menus (see Fig. 6.4 left) are scheduled by the server to be executed on distributed computational resources. In this case, image processing is applied to the entire image collection and is classified as an off-line computation. On the other hand, computations launched inside of the Deep Zoom viewer (see filtering options in Fig. 6.4 right) are executed on a client computer. Client-side image processing is only applied to images viewed in the browser. For the most client computers that are running web browsers, these image filtering computations are interactive. When images are viewed in the Deep Zoom viewer, the web server processes all requests for image tiles and sends them to a web browser as users are panning and zooming. These computations are handled by the web server and are typically interactive depending on the client-server connectivity.

6.3.2   Estimating and Measuring Execution Time

Definition of execution time

To meet execution time requirements, we need to compare multiple acceleration strategies based on a performance metric. We describe execution time and relative speedup as the two main performance metrics. The execution time is derived from the number of clock cycles per execution divided by the clock rate of a processor. For instance, if a program execution takes one million clock cycles on a processor with 1 MHz clock rate, then the execution time is 1 s. The speedup is derived as a ratio of the execution time and the reference execution time for a given computation. One could improve the speedup by using parallel programming models and more computational resources. We divide approaches to obtaining benchmarks for the two metrics into estimation and measurement categories.

Estimation of execution time

Estimation of the execution time for a given task and its algorithmic implementation is possible by evaluating the number of clock cycles per program execution. The difficulty lies in the mapping of complex program computations into clock cycles and then including other contributions to the overall execution time for the entire end-to-end system. Such evaluations almost always involve many approximations. Nevertheless, according to computational complexity theory, computational problems can be classified into classes based on the estimated orders of the number of clock cycles per program execution. For example, if a program must process n image pixels and the number of clock cycles per pixel is 4, then the total number of clock cycles is 4n. This linear computational complexity as a function of n inputs is denoted as O(n) in big O notation. The big O notation hides constant factors and smaller terms. It is very useful for classifying problem independently of hardware specifications.

Estimation of speedup

Relative speedup measurements might be an alternative estimate of interest. Given the cost of advanced hardware and the amount of time spent writing parallel programs, one may want to predict expected computational speedups as a function of the needed investments of time and money. In the case of WIPP, the speedup for a deployment on a computer cluster or cloud can be predicted using Amdahl's law [2]. The speedup S is defined as a ratio of the execution time on a single machine $T(1)$ over the execution time on P processors being utilized in the cluster $T(P)$ as presented in equation below:

$$S(P) = \frac{T(1)}{T(P)} = \frac{1}{\alpha + \frac{1}{P}(1-\alpha)} \tag{6.1}$$

where α is the nonparallelizable fraction of an algorithm. Amdahl's law assumes that the input data size and the amount of computation are fixed. This is typically the case of uploading an image collection from one experiment and processing it on a deployed instance of WIPP.

It is possible to redesign algorithms in WIPP as computational hardware becomes faster. In this case, Gustafson's law [3] can be used instead of Amdahl's law to fully exploit the improving computing power over an increasing input data size in a fixed execution time. Gustafson's law is shown below with the same notation as above:

$$S(P) = \alpha + P(1-\alpha) \tag{6.2}$$

Measurements of execution time

Finally, one can collect actual execution time benchmarks on a specific hardware and software configuration. There are three types of time measurements:

1. Wall clock time: the observed time elapsed between the start and the end of the program measured by an external clock
2. User CPU time: the total time used by the computer's processor executing just the code of the user's program

3. System CPU time: the total time used by the processor executing kernel code
   (i.e., the core of an operating system) on behalf of the program

The kernel code is called from a program, for instance, when read or write operations are performed (also called the system calls). In a case of parallel computations, wall clock time is usually less than user CPU time because the program is run concurrently with other programs and must also be waiting for disk, network, or other devices.

Practical notes about execution time and speed-up metrics

Absolute execution time measurements are typically obtained as an average over a set of repeated runs, which are needed because of varying background processes running concurrently with the measured program. The disadvantage of absolute measurements is that they are hard to use for predicting execution times with different hardware and software configurations. If these configurations remain constant and are replicated across the multiple computational nodes of a cluster, then one can collect speedup benchmarks and use them for predictions. These benchmarks capture an execution time as a function of the number of nodes. They provide better understanding of computational scalability and are useful for trade-off decisions between the shortest execution time and the minimum cost of computational nodes. If both software and hardware configurations across all computational nodes are not the same (e.g., heterogeneous computer cloud), then the speedup benchmarks correspond to the ratio of the worst-case execution time of the fastest sequential algorithm on one of the nodes to the worst-case execution time of the parallel algorithm on all the computational nodes.

Remark

While we have focused on execution time, we omitted the discussion about the amount of RAM required by a program (or space complexity). This type of analysis is important in the case of big image data and must be considered when choosing a parallel programming models and a hardware architecture. Although it is recommended to collect memory benchmarks while writing a program, the space complexity estimation and RAM consumption measurements were not included in the scope of this book.

6.4   From Commercial Big Data Analytics to Research Big Image Analyses

There is a wealth of knowledge gained from building commercial big data analytic solutions that could be leveraged when designing big microscopy image analytic solutions. To learn from them, one can take the following steps:

- Narrow down big data attributes (4 to 7Vs) in commercial applications to those in microscopy imaging laboratories.
- Extract basic and advanced design considerations.
- Apply the design considerations to the design of big microscopy image analyses.

We describe these steps in the rest of this section.

Microscopy image attributes

For microscopy imaging laboratories, image collections are typically of the order of terabytes with velocity about 100 MB/s, and variety is represented by file formats, imaging instruments, and imaged specimens. Veracity is present in microscopy images due to manually selected microscope settings and many calibration protocols. The key characteristic of images is that they have always spatial grid structure as opposed to unstructured data (ignoring for now image annotations) occurring in many commercial big data sets.

Basic design considerations

As commercial solutions address big data analytics for all big data attributes, basic general design considerations can be observed in all big data solutions.

- Solutions must be modular in terms of hardware and software because data attributes, algorithms, and hardware specifications change all the time and modules must be replaced/upgraded (i.e., survivability).
- Software must utilize hardware to its maximum but also must handle hardware failures (i.e., utilization and profit including redundancy and reliability).
- Solutions should support creating processing workflows (i.e., flexibility via functional reconfiguration).
- Data must have identifiers, immutability, and introspection (i.e., data persistence. The data elements are found using unique identifiers, are stored in perpetuity, and can describe themselves in terms of content and relationships [4]).
- User interfaces to installation and operation aim at "zero installation time" and "zero user interface" (i.e., minimum barrier for users).

Additional design considerations

Beyond the basic consideration, big data analytic solutions must also have:

(a) Data access management and tools for data de-identification for information privacy
(b) Data format standards and tools for format transformations (legacy data)
(c) Data quality and tools for data cleaning
(d) Data reduction and tools for such transformations
(e) Performance verification and the tools for integrity of data and correctness of functionalities
(f) Software and hardware interoperability and the tools for verifying interoperability of replaced components
(g) Data preservation

Depending on application-specific requirements, these considerations should be included in a solution design.

Applying basic design considerations

WIPP has incorporated some of these design considerations. The software consists of modules, such as the Pegasus scientific workflow for integrating algorithms and HTCondor for utilizing multiple computers. Each image collection, intermediate data product, or computational job is associated with a unique identifier. Once an input collection is locked before computation, it becomes immutable. A simple query to a database provides information about any collection or executed job. To meet the "zero installation time," a Docker container is used for packaging and deploying the software (and Docker swarm on multiple machines). The "zero user interface" requires more inputs from a community of users and has been implemented so far for the traditional mouse and keyboard devices. Future modifications to WIPP will also address specific additional design considerations.

Incorporating a spectrum of application-specific hardware and software considerations is not trivial. We selected three parts of big image analytic solutions that are of concern to users, algorithmic contributors, and web system developers:

1. *Human interface*: how to interface human inputs and interactions with the output of a big image data solution
2. *Storage and data structures for big images*: how to organize and store large volumes of complex image data
3. *Parallel computations*: how to break image computations into task- and data-parallel components

We will focus in the rest of this chapter on these key parts of a client-server solution for processing big images.

6.5   Human Interfaces for Big Image Data Analytics

Spectrum of User Interfaces
We start with the human interface to big data solutions because humans are the most important part of any scientific discovery. Users come with different levels of IT knowledge and experience with software tools. They also pursue multiple goals by executing a sequence of computations. Depending on the user's knowledge, experience background, and goals, user interface (UI) requirements for big data solutions might include:

(a) Predefined menus and buttons for configuring and executing computations
(b) Scripting and plugin templates for automating computations
(c) Application programming interfaces for integrating new functionality
(d) Application programming interfaces for replacing or adding modules to the entire system (i.e., image processing module, feature extraction module, or machine learning module)

The above UIs can also be classified as:

1. Graphical user interfaces (GUIs)
2. Command-line interfaces (CLI)
3. Application programming interfaces (APIs)

The large variability in user interface requirements implies that a big data solution cannot just have one type of interface for all users.

User interfaces in client-server systems

In client-server systems, the user interfaces (UIs) are on both client and browser sides. We will focus only on a client-side GUI consisting of predefined menus and buttons for configuring and executing computations. The client-side GUIs can be customized by researchers who are knowledgeable about HTML5, CSS, and JavaScript. These UIs are of interest since the reader is assumed to be interested with the easy-to-use aspects of web systems like WIPP. We also provide an example of the GUI design process for the web statistical modeling tool.

6.5.1 Focus on Client-Side Graphical User Interfaces

GUI elements

The objective of GUI design is to present interface elements that are easy to use, access, and understand to facilitate the above activities. The interface elements can be classified as follows[5]:

1. Input controls (e.g., execution launch via buttons; selection via radio buttons, checkboxes, drop-down lists).
2. Navigational components (e.g., page navigation via breadcrumb or pagination, search via search field, sequence navigation via slider or icons).
3. Informational components (e.g., short description via tooltips or modal windows, status of computation via progress bar and icons, warning and error reports via notifications or message boxes).
4. Containers (e.g., toggle between hiding and showing multiple functionalities or large amount of content: via accordion in JavaScript, encapsulating image and processing functionality via JFrame using Java Swing library).

To meet the objectives of GUI design, these elements must be integrated following concepts from *interaction design* focused on interactive digital products and services, *visual design* concentrated on print or electronic forms of visual information, and *information architecture* concerned with organizing and labeling online sites and software to support usability and findability.

GUI design

GUI design anticipates user intentions. For WIPP, the client-side GUI assumes that users intend to:

(a) Uploading and downloading data
(b) Searching for image collections and other data types
(c) Selecting and configuring computations to launch on a server
(d) Browsing results of computations
(e) Viewing big images
(f) Selecting and configuring computations to launch on a client while viewing big images

---

[5] https://www.usability.gov/what-and-why/user-interface-design.html

With a list of anticipated use cases, best practices for designing GUI can be put in place to address simplicity, consistency, use of color and texture, layout, and typography to assure legibility and readability.

6.5.2   Example of GUI Design for web Statistical Modeling Tool

GUI design for web statistical modeling tool

Let us consider a GUI design for the web statistical modeling tool described in Chap. 2, Sect. 2.4. A user wants to derive a statistical probability distribution function (PDF) model from cell colonies that have been segmented from a sequence of gigapixel images and described by a set of cell colony measurements (features). The user interface should allow the following actions:

1. Selecting an imaging channel
2. Selecting a colony feature
3. Defining number of histogram bins
4. Filtering data considered for modeling by their spatial location
5. Filtering data considered for modeling by their feature value
6. Computing and showing statistics of selected and filtered data
7. Suggesting PDF model type
8. Computing and showing PDF model parameters
9. Saving the final histogram with traceable hyperlinks for each cell colony to its persistent source

In this example, the functionality is divided into parameter selection (1–3), spatial filtering (4), feature filtering (5), statistical modeling (6–8), and publication (9). The GUI design for parameter selection is implemented using input controls, such as drop-down menus, an edit box, and a spinner. The rest of the functionality is encapsulated in an accordion type of a container (see Fig. 6.5). Within the accordion, all statistical modeling and publication functions are launched using input controls, such as buttons with icons, which are followed by informational components, such as message boxes, images, and tooltips. In contrast, all filtering functions are using either slider bars for feature values or an image viewer for cell colony locations shown with color-coded markers based on a feature value.

General challenges in GUI design

Perhaps, one of the most general challenges in GUI design is the limited size of device displays. Depending on the stage of user's activities (selecting parameters→ filtering→statistical modeling), different information is more (or less) important to users for making their decisions. Thus, the GUI could reallocate the use of display size depending on the activity which was achieved by an accordion element (hiding and showing input controls) in the example above. This concept has also been implemented in the design of integrated development environments (IDEs) where

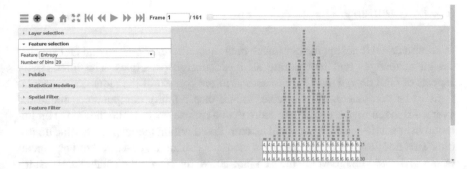

Fig. 6.5 GUI of web statistical modeling tool with the accordion type of a container (left) and a canvas for display (right)

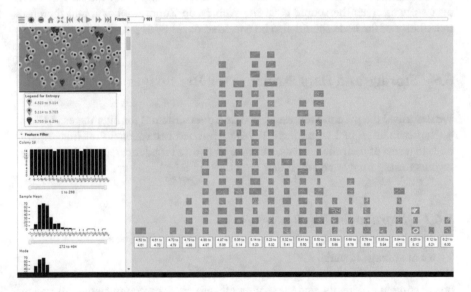

Fig. 6.6 Filtering challenges in the web statistical modeling tool in terms of display size

switching, for example, between programming, debugging, and searching activities, triggers new layouts.

Specific challenges in GUI design for big image data

Specific challenges in big image data arise when showing information for spatial and feature filtering since there is no display size that could accommodate gigapixel images and thousands of feature histograms with sliders (see Fig. 6.6). For spatial filtering, one must adopt multi-resolution representations of gigapixel images to enable pan and zoom. For feature filtering, one can use a scroll bar to view feature histograms beyond those that fit on a finite display.

6.5.3   Summary

In summary, GUI design for big image data must address both general and specific design challenges and incorporate all basic design principles. The interactivity aspects of GUI design must be understood in the context of the requested computations to be completed. For example, image thresholding computations requested over a TB-sized image might take more than a mouse click, while the same computation over a MB-sized image could be completed within the interactive time definition. This implies that a GUI design for visual optimization of a threshold parameter over a MB-sized image would have an interactive interface (click and render result). In comparison, a GUI design for the same computation over a TB-sized image would have an interface that consists of unique identifiers to check the status of the computation completion (i.e., status message or hour glass per unique identifier) and an interface to retrieve its results. Finally, a GUI design can become very complex software (see the source code for Web Deep Zoom Toolkit[6]), and therefore modularity of the code should also be considered.

6.6   Storage and Data Structure for Big Images

We described the pyramid representation as a data structure for big images in Chap. 4 (Representation of Large Images). We mentioned heterogeneity of image pyramids in terms of their file formats. Here we provide a broader perspective on storage layouts for big images and their data structures in RAM.

6.6.1   Storage for Big Images

Types of storage layout

We are concerned with storing a very large image on a disk, reading and writing its image content efficiently, and preserving all information accompanying all images acquired by a microscope and all information generated during image processing. To achieve maximum performance with big images, one must understand multiple types of storage layouts and their impact on the end-to-end execution time. Figure 6.7 illustrates (1) disk, (2) file storage, (3) image pixel, and (4) pixel byte layout options. We elaborate next on each storage layout option.

Disk storage layout

Big images can be stored on disk(s) in:

1. A single file
2. Multiple files stored as a set of folders on a file system (i.e., the multi-resolution pyramid representation)
3. A database

---

[6] https://github.com/usnistgov/WebDeepZoomToolkit

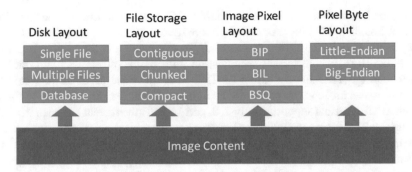

Fig. 6.7 Options of storage layouts for image content. The image pixel layout abbreviations stand for Band interleaved by pixel (BIP), Band interleaved by line (BIL), and Band sequential (BSQ)

In any big image experiment, it is very likely that at least two of these storage types will be used. The reason lies with the fact that to the best of our knowledge, all microscopes store acquired FOVs in a single file. If the acquired files are preprocessed for visualization on the web, then they are likely stored on a file system or in a container file format. If the acquired files are analyzed, then they are likely stored in a database with the image measurements. It is also very likely that an acquired image will change not only its storage type but also its file format at some time point. This is due to (1) a large spectrum of single file formats for storing images acquired by a variety of microscopes and (2) the multiplicity of visualization and analytical goals that an image might support during discovery. Preserving all information during these storage and format conversions is very important for traceability and reproducibility of imaging experiments.

Single file storage
The most commonly used single file formats in microscopy are Adobe TIFF (Tagged Image File Format) and HDF5 (hierarchical definition file, version 5). Both formats have open-source implementations of libraries for reading and writing files. The TIFF specification is described in the 6.0 specification [5], and its implementation in C programming language is available online.[7] The HDF5 specification and its implementation in C programming language are maintained by the HDF group.[8]

TIFF (single file storage)
For files in TIFF format, the topic of preserving information in a single file has been addressed by converting TIFF files into a standard Open Microscopy Environment (OME) representation (see Chap. 5, Loading Images Using OME Bio-Formats Library). The TIFF files acquired by a microscope are stored in OME-TIFF[9] format using the Bio-formats library[10] developed under the umbrella of the Open Microscopy Environment consortium. The topic of TIFF storage size has been

---

[7] http://libtiff.org/

[8] https://support.hdfgroup.org/products/

[9] http://www.openmicroscopy.org/site/support/ome-model/ome-tiff/specification.html

[10] http://www.openmicroscopy.org/site/support/bio-formats5.4/

recognized as an issue in the past because the format has a limit on the stored file size of less than 4 Gibibyte or GiB (4.294967296×10^9 bytes). This limit is due to the 32-bit offset in the TIFF file formats.[11] To overcome this limit of TIFF file formats, the BigTIFF file format specification with 64-bit offsets was introduced in 2007, and the TIFF library called LibTIFF was upgraded as of its version 4.0.[12] Similar issues arose in the past with the number of bits per pixel (BPP). While the original TIFF format supported only 2, 8, and 16 BPP, the format has been extended to support 32 BPP based on the need of many geographical mapping agencies using geospatial information systems (GIS). Although the LibTIFF library (version 4.0) has the support for 64-bit offsets and 32 BPP, many image processing packages still contain older versions of LibTIFF and hence have only support for the 4 GiB limited TIFF format and the pixel depth up to 16 BPP.

HDF5 (single file storage)
For files in HDF5 format, information associated with images is preserved by adding the OME metadata to the HDF5 container (e.g., by using the Bio-Formats library). Due to the popularity of HDF5 for storing very large files, research and commercial communities have built several programming and scripting interfaces to HDF, for instance, API from Java, Python, R, Fortran, Imaris,[13] or MATLAB.[14] The HDF5 format comes with a data model and a library as described in [6]. In terms of HDF file storage size, there is no limit.[15] However, there is a limit of 32 dimension dataspaces (number of bands or channels). In terms of reading and writing speed, HDF5 format contains serial and parallel HDF5 code, and the choice is selected when building HDF5. The key feature of parallel HDF5 is that certain groups of functions must be called collectively if the data value on a target storage node will be modified. Parallel HDF5 can be optimized to maximize the access to image content. The optimization of parallel HDF5 includes setting HDF5 parameters (e.g., chunk size and dimensions), MPI I/O parameters (e.g., the block size and the number of target nodes to be used for collective buffering file access), and parallel file system parameters (e.g., the size of the striping unit and the number of I/O devices to stripe across).

Storage in a set of folders on a file system
In the case of a single file storage, some file formats can store multiple images and/ or image chunks. In contrast, storage of multiple images in a set of well-organized folders might be sometimes more efficient. For example, image pyramid representation can be stored in a set of folders labeled by the pyramid level as illustrated in Fig. 6.8. When accessing image content, the file path can be automatically generated according to the zoom level that corresponds to the folder name. For instance, the folder name 0 refers to low magnification images, and the name 13 refers to high magnification images in Fig. 6.8. The files are consistently named per their grid

---

[11] https://en.wikipedia.org/wiki/TIFF

[12] http://www.simplesystems.org/libtiff/

[13] http://www.bitplane.com/imaris

[14] https://www.mathworks.com/

[15] https://support.hdfgroup.org/HDF5/faq/limits.html

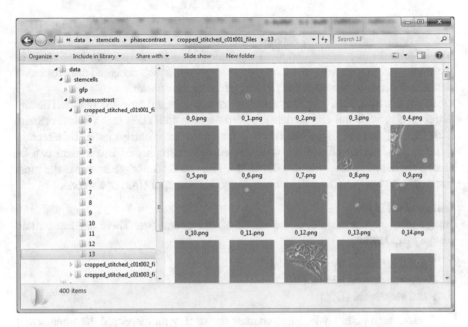

Fig. 6.8 Example of pyramid representation storage in a set of folders on a file system

position (e.g., 0_3.png corresponds to 0th row index and 3rd column index) and hence can be retrieved from the file system efficiently. In this case, we've used the file png format because it is supported by the majority of web browsers.[16]

Database storage

The most common solution for storing big images is to use a file system for storing image chunks and a database for storing indices pointing to the image chunks. Other solutions depend on the size of image chunks, the read/write ratio, and the disk/memory ratio relative to the read/write ratio. If both, large image chunks and their indices, are stored within a database, then the database introduces an overhead by compacting the index look-up table (requires movement of data), accessing data from a fragmented index table (requires inline storage of image chunks in a base table row), and creating many replicates (e.g., log files and database entries, redundancy of immutable data in Hadoop Distributed File System).

File storage layout

The storage layout defines how the pixel values are physically stored on disk. We use the TIFF and HDF5 file formats to explain multiple file storage layouts. For example, HDF5 supports three storage layouts:

1. Contiguous
2. Chunked
3. Compact[17]

---

[16] https://en.wikipedia.org/wiki/Comparison_of_web_browsers
[17] https://support.hdfgroup.org/HDF5/Tutor/layout.html

Contiguous refers to the pixel values stored in one contiguous block of the HDF5 file. In comparison, chunked denotes pixels being stored in equal-sized contiguous blocks (chunks of a predefined size) together with a chunk index to keep track of their association with a dataset. Compact storage layout was designed only for small datasets that can be stored in the HDF5 header of the dataset.

In comparison, TIFF supports contiguous and chunked storage layouts. The key difference is with the chunked layout type where the TIFF tags (RowsPerStrip, StripOffsets, and StripByteCounts) encode separate image strips rather than rectangular image blocks. Otherwise, compressed or uncompressed image data can be stored almost anywhere within a TIFF file. The chunked storage layout is the most important layout type for efficient big image input/output (I/O) operations.

Image pixel layout

Image pixel layout can affect the speed of I/O operations. There are three main image pixel layouts:

1. Band interleaved by pixel (BIP)
2. Band interleaved by line (BIL)
3. Band sequential (BSQ)

These image pixel layouts are schemes for storing multispectral 2D images and their pixel values in a file or in a memory. The three layouts are illustrated in Figs. 6.9, 6.10, and 6.11.

The choice of an image pixel layout depends on the expected image manipulations. For example, the BIP layout is optimal for computing a weighted sum of spectral values at each pixel (spectral analysis). The BSQ scheme is optimal for performing spatial filtering on a single spectral band (spatial analysis). The BIL layout can be viewed as a compromise format for easy access to both spatial and spectral information. All image pixel layouts can accommodate any number of spectral bands.

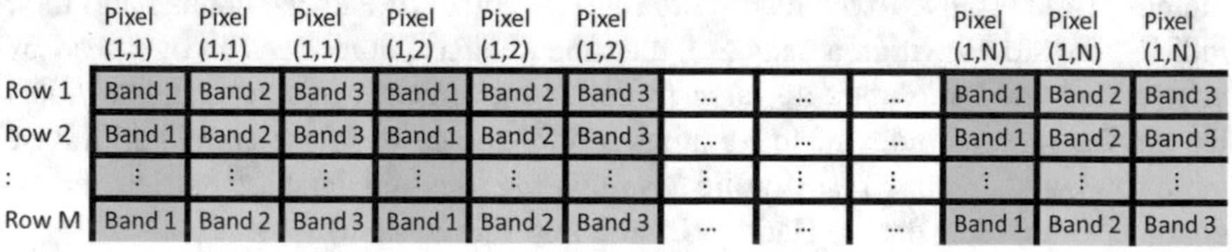

	Pixel (1,1)	Pixel (1,1)	Pixel (1,1)	Pixel (1,2)	Pixel (1,2)	Pixel (1,2)				Pixel (1,N)	Pixel (1,N)	Pixel (1,N)
Row 1	Band 1	Band 2	Band 3	Band 1	Band 2	Band 3	...	...	...	Band 1	Band 2	Band 3
Row 2	Band 1	Band 2	Band 3	Band 1	Band 2	Band 3	...	...	...	Band 1	Band 2	Band 3
:	:	:	:	:	:	:	:	:	:	:	:	:
Row M	Band 1	Band 2	Band 3	Band 1	Band 2	Band 3	...	...	...	Band 1	Band 2	Band 3

Fig. 6.9 Band interleaved by pixel (BIP) layout of image pixels

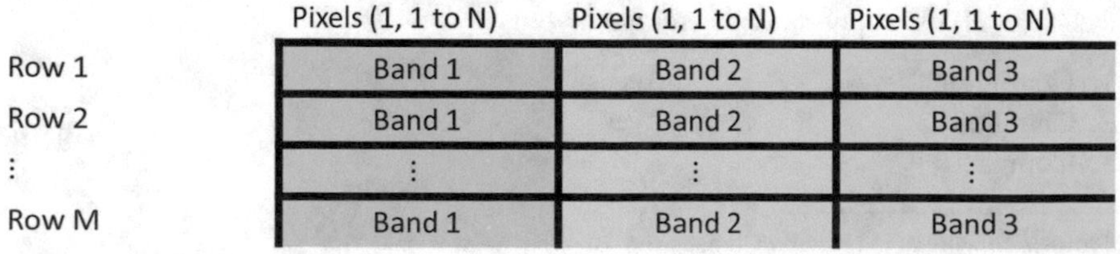

	Pixels (1, 1 to N)	Pixels (1, 1 to N)	Pixels (1, 1 to N)
Row 1	Band 1	Band 2	Band 3
Row 2	Band 1	Band 2	Band 3
:	:	:	:
Row M	Band 1	Band 2	Band 3

Fig. 6.10 Band interleaved by line (BIL) layout of image pixels

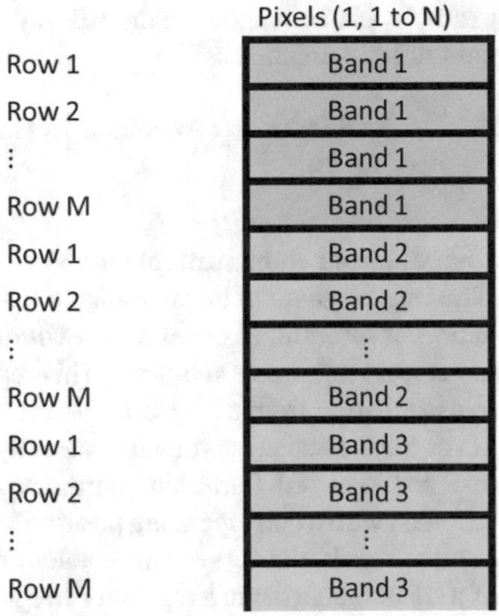

Fig. 6.11 Band sequential (BSQ) layout of image pixels

Pixels (1, 1 to N)

Row 1	Band 1
Row 2	Band 1
⋮	Band 1
Row M	Band 1
Row 1	Band 2
Row 2	Band 2
⋮	⋮
Row M	Band 2
Row 1	Band 3
Row 2	Band 3
⋮	⋮
Row M	Band 3

Pixel byte layout

Each pixel is associated with a value represented by 2, 8, 16, 32, or 64 bits per pixels. During network transmission, byte order is important for values represented by more than 8 bits (1 byte = 8 bits). There are two sequential byte orders: little-endian (least significant bits first) and big-endian (most significant bits first). For example, the number 5 would be binary-encoded using 16 BPP as 0000 0101 0000 0000 using big-endian and 0000 0000 0000 0101 as little-endian. The two sequential orders have implications on storage and are encoded in the file's metadata section.

6.6.2 Data Structures for Big Images

Data structures are the representations of images in RAM. Due to the finite size of RAM, we categorize big image data structures based on the ratio of RAM size to image size. We will briefly mention images with more than two dimensions (denoted as 2D+) and their pyramid representations. WIPP supports 2D+ image collections of videos and spectral sequences with gigapixel 2D images (i.e., XY + time or $XY + \lambda$).

Images smaller than available RAM

If an image can fit into RAM, then it can be represented as a multidimensional array of values. Because many image operations require accessing all the pixels using programming language looping constructs, a one-dimensional array is an efficient representation for high-dimensional images. In the body of the loop, subsequent pixels are accessed by incrementing an index, which is more efficient than calculating the position of the next pixel from the row, column, and band values. For example, if the image is in the BIP layout, then a pixel value at the location [row, column] is computed as

$$\text{pixel index at}\left[\text{row,column}\right] = \left(\text{number of columns} \times \text{row} + \text{column}\right)$$
$$\times \text{number of bands} + \text{band} \qquad (6.3)$$

Images larger than available RAM

If an image cannot fit in the available RAM, then it must be represented as a combination of multidimensional arrays with the indices defining location in relative to the current big image sub-array. This representation can be viewed as a coordinate transformation from a 2D Cartesian system (i.e., big image rows and columns) to a 4D or 5D Cartesian coordinate system that enables access to image subareas that fit in RAM. The 4D Cartesian coordinate system consists of a set of small images (chunks) with a corresponding position vector for their coordinates in the big image. In other words, the 4D system contains one set of 2D coordinates giving the position of a small image tile in a regular or irregular grid and the other set of 2D coordinates referring to a pixel position inside of a small image tile. The 5D Cartesian coordinate system consists of a multi-resolution pyramid of small images (tiles). It can be viewed as a location in a stack of 4D Cartesian coordinate systems (or a 5D pyramid) where the fifth dimension is the resolution. This representation has been proven to be very efficient for big 2D images.

2D+ images

In practice, microscopy images are not more than two-dimensional. However, microscopy experiments over a large FOV can generate time-lapse images (e.g., XY + time videos from phase contrast microscopes), high-dimensional spectral images (e.g., $XY + \lambda$ data from coherent anti-Stokes Raman microscope or scanning electron microscope with energy-dispersive X-ray spectrometry), confocal z-stack images (e.g., XYZ data from confocal laser scanning microscope), or a combination of time-lapse, spectral, and z-stack images (e.g., $XYZ + \lambda$ data from confocal z-stack images with more than three fluorescent channels or XYZ + time video from confocal laser scanning microscope).

Pyramids for 2D+ images

We can view these high-dimensional terapixel images as a set of 2D cross-sectional images. If we decompose 3D+ images into 2D cross sections, then we can use a pyramid representation suitable for big 2D images. These 2D cross sections can be represented as an ordered set of multi-resolution pyramids using one pyramid per 2D cross section with the order defined by the third dimension. To enable fast rendering and processing, one may have to generate three sets of pyramids that correspond to the three orthogonal 2D cross sections per one large 3D volume (image). For example, for a 3D image with XY + time dimensions, one would generate sets of pyramids for $\{XY\}$, $\{X + \text{time}\}$, and $\{Y + \text{time}\}$. This representation works well for XY + time or $XY + \lambda$ images where oblique views are not meaningful for visual inspection. However, this representation might be limiting for 3D images with XYZ dimension images because the oblique views are important for data explorations.

6.6.3   Summary

In the design of a big data analytic solution, there are many decisions about big image storage and representation that determine image content access efficiency. It is possible to optimize the design for a given set of big image analyses with a well-defined pattern of accessing image content. Given hardware specifications (RAM, disk, and bus), optimal parameters can be determined in terms of pixel byte, image, file, and disk storage layouts as well as data structures. In practice, it can be difficult to predict image content access patterns and anticipate the hardware used. While this unpredictability can explain suboptimal image analytic software performance, it also highlights the importance of considering application requirements and usage patterns when making software design decisions.

6.7   Parallel Computations Over Big Image Data

Software development for big data must address three problems:

1. Algorithmic design to automate processing
2. Integration of heterogeneous algorithms into software systems to leverage existing software investments
3. Algorithmic implementations that integrate software and hardware

The last problem of integrating software with distributed hardware resources is the topic of parallel computing research. We provide a high-level classification of parallel programming models and briefly describe each model.

Parallel computing
The basic premise for accelerating big data image computations is that (a) the images can be divided into smaller tiles and (b) the computations can be divided into smaller functional tasks that are then applied in parallel to the smaller image tiles (image data and functional decompositions). Parallel execution accelerates calculations by utilizing multiple computational resources but comes with the cost of additional hardware resources and of writing the software specific to a hardware architecture.

Classification of parallel programming models
The several commonly used parallel programming models abstract hardware and memory architectures and provide different programming approaches. Parallel programming models[18] that are derived from computer architectures include:

1. Shared memory model (without or with threads)
2. Distributed memory model with Message Passing Interface (MPI)
3. Partitioned Global Address Space (PGAS) model with data parallel decomposition
4. Hybrids of the above

---

[18] https://computing.llnl.gov/tutorials/parallel_comp/#Whatis

Parallel programming models can be divided into two broad categories based on their structural granularity of their parallel programs:

1. High-level granularity
2. Algorithmic-level granularity

These categories can be further subdivided. Those with high-level granularity have been divided into:

1. Single program multiple data (SPMD) model
2. Multiple program multiple data (MPMD) model

while those with algorithmic-level granularity[19] can be further classified as:

1. Data parallel model
2. Master-agent model
3. Task graph model
4. Task pool model
5. Producer-consumer model
6. Hybrid model

These categories might be overwhelmingly complex for a WIPP user and somewhat complex even for a developer of WIPP algorithms since they require basic understanding of hardware and software.

In this chapter, our approach is to introduce the WIPP developers of algorithms to parallel programming models at the algorithmic-level granularity and incorporate their knowledge about the hardware used for running WIPP in their algorithmic design. We assume that a reader is familiar with a hardware architecture running WIPP which typically includes RAM, CPUs, communication buses for exchanging data, and pluggable graphics processing units (GPUs). Next, we will briefly describe each of the algorithmic-level models applicable to such hardware architectures in the context of image processing.

6.7.1   Data Parallel Model

This model is based on dividing images into smaller regions and applying the same processing to each region on a separate hardware resource (i.e., a computational node). The most difficult aspect is determining the image partition strategy that will colocate each computation with its needed data [7]. To illustrate the difficulty, we list a few spatial image computations in Table 6.1 as examples motivating different partition strategies. For instance, if computations operate on a single pixel (e.g., thresholding that has no spatial overlap with the computation of a neighboring pixel), then image partition can be based on either physical location of a pixel in a file or logical location of a pixel in an image. However, if computations operate on

---

[19] https://www.tutorialspoint.com/parallel_algorithm/parallel_algorithm_models.htm

Table 6.1 Subdivision of spatial image computations and its relevance to image partition

Types of spatial computations: examples	Input image region	Overlap type	Desired image partition	Input to logical partition
Pixel-based: Thresholding	Fixed size	No overlap	Physical or logical without overlap	None
Kernel-based: Convolution	Fixed size	With overlap	Logical with overlap	Kernel area size
Segment-based: Feature extraction	Variable size	No overlap	Logical without overlap	Mask
Bounding box based: Background correction	Variable size	With overlap	Logical with overlap	Bounding boxes

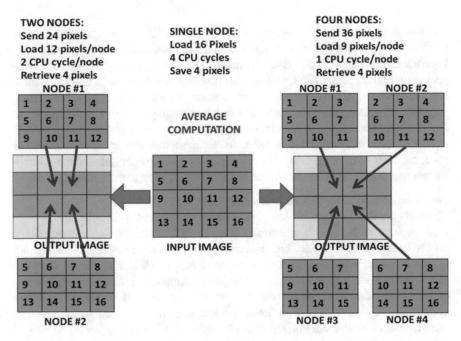

Fig. 6.12 Examples of image partitioning for spatial averaging over 3 × 3 pixels for one (middle), two (left), and four (right) distributed computational nodes

a pixel neighborhood (e.g., kernel-based), then the image partition strategy should be based on logical location of pixels.

Example

As a specific example, Fig. 6.12 numerically evaluates the advantages of logical partitioning for the case of spatial averaging over 3 × 3 pixels with overlapping four image regions. The computation is illustrated for a number of computational nodes N equal to one (middle), two (left), and four (right). The comparison across an increasing number of nodes shows how the decreasing numbers of CPU cycles and the pixel load operations to RAM are counterweighted by an increasing number of communications (send and retrieve operations).

Practical notes

The partitioning can be performed by horizontal and vertical image cuts. To avoid any exchange of pixels between nodes during runtime, the cuts are made in such a way that there is redundancy while spraying pixels across computational nodes. The goal is (a) to avoid the communication overhead between nodes and the storage area network (SAN) or network-attached storage (NAS) where the data is stored and (b) to mitigate the impact of node failure rates of large computer clusters.

6.7.2   Master-Agent Model

The master-agent model is based on introducing a hierarchy of computational nodes to divide the work into (a) "generating" computational jobs and (b) executing the jobs. Figure 6.13 illustrates the master-agent model. The word "generate" refers to decomposing a workflow to individual tasks, collecting data needed for each task in a case of distributed memory, and assigning and transmitting data to agents. One or more master processes manage the computational jobs by delegating them to agent processes so that they are executed in the shortest time, a difficult process known as "load balancing." The load-balancing strategy can consider more than the agent utilization. For instance, other potential factors to consider are data throughput times, agent reliability, past agent execution times, and the patterns of incoming computational tasks.

Job scheduler to execute load balancing

The software that implements load balancing is called a job scheduler or a distributed resource manager. Job scheduling is concerned with assigning computational jobs to computational nodes which is distinct from operating system process scheduling concerned with assigning running processes to CPUs. The input to a job scheduler is a job queue which contains job information. Job schedulers are frequently embedded in scientific workflow systems, such as the Pegasus workflow system used by WIPP. The workflow for a sequence of computations also defines task dependencies that are utilized in master-agent models and in job schedulers.

Hadoop example of a master-agent model

An example of the master-agent model is the Apache Hadoop framework,[20] which is designed for storing data and running big data computations on computer clusters and clouds consisting of commodity hardware. Data storage is supported by the Hadoop Distributed Filesystem (HDFS) which uses the master-agent model. In this model, one cluster node (labeled as NameNode) manages file system operations, and a set of agents (labeled as DataNodes) manages data storage on their individual cluster nodes. When NameNode sprays data blocks across DataNodes, the blocks are replicated multiple times. The defaults are 64 MB blocks with two replicates. If a DataNodes fails, then the NameNode finds the replicates elsewhere in the cluster,

---

[20] https://hadoop.apache.org/docs/r2.7.1/hadoop-mapreduce-client/hadoop-mapreduce-client-core/MapReduceTutorial.html#Mapper

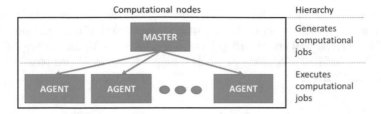

Fig. 6.13 Master-agent model and its hierarchy of computational nodes

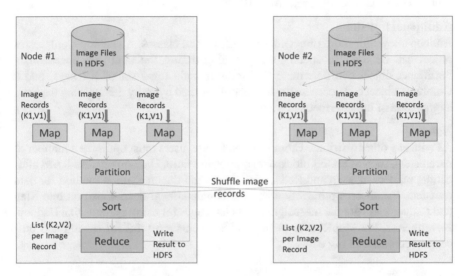

Fig. 6.14 Simple illustration of MapReduce programming paradigm utilizing two computational nodes

allowing the computation to be restarted on a different computational node while the operator replaces the failed hardware.

Map-reduce implementation

Several WIPP image processing algorithms have been evaluated using Hadoop implementations [8]. The computation with Hadoop is based on the MapReduce programming paradigm [1]. We will explain the MapReduce programming paradigm using a simple two-node cluster to perform an image intensity histogram calculation. We start by assuming that image pixels have already been sprayed across HDFS and each node has a subset of the image pixels, as illustrated in Fig. 6.14.

Map function

The image intensity histogram calculation starts with a Map function that computes frequency count for each intensity value over an allocated set of pixels. This Map function can be viewed as a transformation of pairs from one space to another space; the Map transforms a list of pairs (K1 = pixel location, V1 = intensity) to another list of pairs (K2 = intensity value, V2 = frequency count). Once executed, Hadoop will exchange groups of entries between the available nodes as illustrated in Fig. 6.14 by "partition" and "sort" operations along with the Hadoop shuffling operation. In our example, Hadoop will create Group 1 that has K2 in [0, 128] and Group 2 that has K2 in [129, 255]). Then, Node #1 sends Group 2 to Node #2 and Node #2 sends Group 1 to Node #1.

Reduce function
A previously written Reduce function merges the entries with the same K2 value and saves the resulting counts in HDFS. After the Reduce function, the image histogram counts can be retrieved from HDFS.

Additional functions
Hadoop allows for more direct control with several classes. A Hadoop Comparator object can be used to specify the grouping during the shuffling operation, a Partitioner object can determine the Reducer node for a set of K2 keys, and a Combiner object can decrease the number of shuffled bytes by determining the local aggregation of the intermediate Map.

Notes
As with any other distributed computing software, users must optimize a number of parameters, such as HDFS block size, replication ratio, Hadoop process RAM allocation, number of Map and Reduce instances, and the encryption method for data transfers. Parameter optimization and the computational decomposition into Map and Reduce tasks are the most significant challenges for using systems like Hadoop. Nevertheless, the MapReduce paradigm has been successful for storing and processing commercial big data.

6.7.3   Task Graph Model

The task graph model describes a computation as a directed task graph formed by a collection of vertices denoting atomic tasks and directed edges describing data movement. An atomic task is a logically discrete section of computation in a program that is executed by a processor. A task graph-based parallel algorithm consists of atomic tasks running on multiple processors or computational nodes. All computations are executed by traversing the task graph by following the directed edges. A graph is a directed acyclic graph (DAG) if there are no paths that start at a vertex, follow a sequence of directed edges, and return to the starting vertex. Figure 6.15 shows an example of DAG with eight atomic tasks T1–T8. Programs that can be represented by a DAG are characterized by useful properties, for instance, a reachability relationship. In Fig. 6.15, we illustrate the reachability relationship with vertex T8 reachable from the vertex T1 (T1 $\leq$ T8) if there is a path from T1 to T8.

Practical use task graph models
Practical applications of task graphs include task scheduling, dataflow programming, and management of software revision history and its versions. This

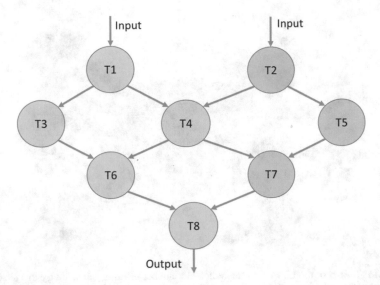

Fig. 6.15 Example of a directed acyclic graph with eight tasks

programming model is recommended for computations where the transfer time of moving data is larger than the total time needed for the number of computations associated with the data tasks. In this case, the task graph structure can be optimized to lower the data movement cost between the tasks. The most difficult aspect of implemented task graphs is the decomposition of computations into atomic tasks, identifying task synchronization and communication dependencies, and the creation of the directed task graph. For example, the color coding of vertices shown in Fig. 6.15 can be used for assigning tasks to three computational resources assuming that all tasks require the same amount of time to complete. The assignment becomes complicated when the computational resources and tasks are heterogeneous in terms of CPU power, RAM, data needs, and computational complexity (i.e., the assignment becomes a load-balancing problem).

6.7.4 Task Pool Model

The task pool model can also be thought of as a type of task graph model. It is based on dynamic assignment of tasks to the computational nodes to balance the load. There is an advantage to creating a pool of tasks when the task completion time is unpredictable or varies significantly. The model consists of the implementations of a master and agent processes. The master generates and holds a pool of tasks, sends tasks to agents upon request, and collects the results. The agents request and receive tasks from the master, execute the tasks, and return the results to master. If the pool of tasks is generated dynamically, then a method of detecting termination is required

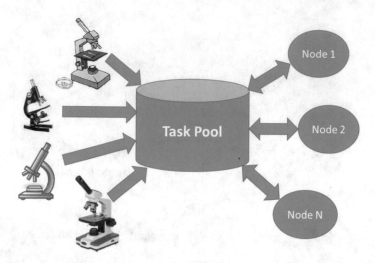

Fig. 6.16 Task pool model with an unpredictable number of tasks generated by image streams from multiple microscopes

so that all agents stop requesting tasks when the supply is exhausted. The number of tasks executed by each computational resource will depend on its speed of execution because there is no preassignment of tasks.

Practical notes
The task pool model is used when the amount of data associated with each task is small and therefore the communication time overhead in sending tasks and receiving results is smaller than the total time needed for computations. One of the difficult design decisions is choosing the task granularity to optimize relationship between the communication overhead, the computational effort, and the data quantities. For example, if the goal is to index acquired images by their average intensity, then a task pool model is appropriate for processing each incoming image from a set of microscopes. As illustrated in Fig. 6.16, the number of images per time unit depends on microscope acquisition rate and its usage pattern which dynamically determines the tasks generated to compute an average image intensity. Computational nodes request the tasks with the time needed to complete a computation depending on the image size. The task pool model inherently balances the computational load despite varying image sizes.

6.7.5   Producer-Consumer Model

The producer-consumer model (also called the pipeline model) represents a computation as a chain of multiple data producers and consumers in a manner similar to an assembly line. Each computational task in the queue consumes data from the preceding task and produces data for the subsequent task. The queue can be linear or

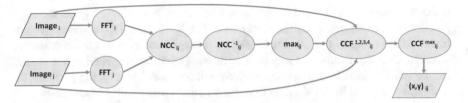

Fig. 6.17 Computation of relative displacements (x, y) of two image tiles (images i and j) using a consumer-producer model

represented by a directed graph. The producer-consumer model is different from a task graph model by overlapping task interactions with computations. Both, producer and consumer, share a common, fixed-size buffer used as a queue. The producer generates data, puts them into the buffer, and starts its task-specific computation again. At the same time, the consumer removes data from their common buffer, one chunk at a time, and starts its task-specific computation. The producer will wait if the buffer is full, and the consumer will wait if the buffer is empty. In other words, every time a producer task generates new data, it triggers the execution of a next consumer task in a queue.

Stitching algorithm
Among several image processing algorithms in WIPP, the tile stitching algorithm has been implemented for multiple architectures using multiple producers and multiple consumers [9]. The tile stitching computation is based on normalized correlation coefficients (NCCs) derived from fast Fourier transforms (FFTs). The image tile displacements (i.e., translation vectors (x, y) for each pair of adjacent tiles) can be computed using multiple data and functional decompositions, as well as CPU and GPU hardware architectures. Figure 6.17 shows the sequence of steps for multi-threaded CPU-only implementation. After each image tile is read into memory and processed by FFT, NCCs are computed. Afterward, 2D inverse FFT of the NCCs is performed ($NCC^{-1}{}_{ij}$), and maximum is found (max_{ij}). The final steps are to compute the cross-correlation factors (CCF) for the north, west, south, and east overlaps, find the maximum CCF, and save the corresponding translation vector (x, y). The key aspect is the assignment and management of threads (one for reading an image tile, one for computing FFT, one for completing NCC, FFT^{-1} and max, and multiple threads for CCF). The difference between simple sequential and pipelined CPU implementations can yield a speedup by almost an order of magnitude and with GPUs even higher [9].

Practical notes
The consumer-producer model has been used for developing general dataflow environments (i.e., computational scenarios where data flow along a processing pipeline). These environments became popular in scientific workflow management systems (e.g., Kepler [10], Taverna [11]) and in commercial frameworks such as the .NET framework (TPL Dataflow Library). A difficulty with the consumer-producer model

is the implementation of multiprocess synchronization. Nevertheless, given the large number of open-source scientific workflows [12] that utilize dataflows, we recommend using one from the existing workflow management system implementations when suitable. For example, a sequence of filtering operations applied to big images can be implemented by using a consumer-producer model to pass the partially filtered image regions in RAM rather than passing them between RAM and disk.

6.7.6 Hybrid Model

The hybrid model combines multiple programming models either hierarchically or sequentially to any part of a computational algorithm. Hybrid models are motivated by integrating functional and data decompositions with specific strengths of a hardware architecture.

Example with CPU-GPU hardware
The integration of functional and data decompositions can be achieved by combining CPUs and GPUs at two different levels. At the low level, the integration takes place by putting CPU and GPU on the same die and sharing the on-chip cache and off-chip memory [13]. At the high level, the integration is accomplished by attaching GPUs to a computer with CPUs and orchestrating the split of computations between CPU and GPU units [14]. In the latter case denoted as a single CPU-GPU configuration, the GPUs are typically assigned data parallel work to take advantage of their large number of computational cores, while the CPUs execute sequential code or data transfer management.

Given a CPU-GPU configuration illustrated in Fig. 6.18, the hardware resources consist of several computational nodes with master CPU and multiple GPUs. Following the classification introduced in Sect. 6.7, the utilization of this hardware configuration can benefit from distributed memory model with Message Passing

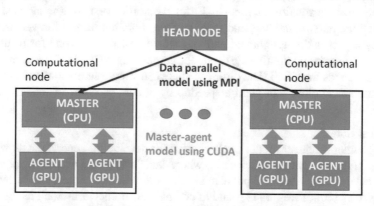

Fig. 6.18 Hybrid parallelization model combining master-agent model using CUDA interface and data parallel model using MPI on a CPU-GPU cluster hardware configuration. CUDA stands for Compute Unified Device Architecture (rarely used in a non-abbreviated form)

Interface (MPI) to run single process, multiple data (SPMD) parallel applications. At the algorithmic-level granularity, one can create a hybrid parallel model consisting of (1) a data parallel model with the MPI protocol (see Chap. 4) for launching tasks on each computational node and (2) a master-agent model for leveraging all GPUs attached to each computational node. In this hybrid programming model, each node receives data from the head node using MPI, exchanges the received data between its local memory and the attached GPUs using GPU-specific interface, and collects computed results. To interface GPUs, one option is to use CUDA®, a parallel computing platform and programming model introduced by NVIDIA.

Practical Notes
In practice, the number of possible hardware configurations is very large, and therefore there is no recipe for creating an optimal hybrid parallel model. Based on the current economics of building computer hardware and providing computer cloud services, developers of algorithms are frequently writing code to run on computer clusters and distributed virtual machines with multi–/many-core machines. In this case, it is beneficial for the developers to be familiar with the parallel programming models described in this section and with the interfaces for direct multi-threaded, shared memory parallelism, such as the Open Multi-Processing (OpenMP[21]) API.

6.7.7   Summary

This entire section focused on parallel computations over big image data and particularly on algorithmic implementations that integrate software and hardware. The algorithmic implementations and approaches followed parallel computing models that were derived from computer architectures and divided based on the structural granularity of their parallel programs. The structural granularity provided a mechanism for classifying parallel programming models into single/multiple program multiple data (SPMD/MPMD) models and a variety of algorithmic-level models (data parallel, master-agent, task graph, task pool, producer-consumer, and hybrid models). The parallel computing premises (data and computation subdivision) and algorithmic-level models were presented at a high level in order to introduce a reader to writing algorithms that leverage hardware.

Forward-looking challenges in algorithmic programming
With the increasing variety of hardware architectures, the parallel programming models are still an open research area. The hardware architectures are changing not only in terms of scale and density (e.g., 5 billion transistors per die, feature sizes close to 10 nm) but also in terms of brand new structures, such as neuromorphic computing and quantum computing. For example, in the early 1990s, scientists began considering a design of brain-like (neuromorphic) computing devices that would dramatically outperform conventional Complementary Metal–Oxide–Semiconductor (CMOS)

---

[21] https://computing.llnl.gov/tutorials/openMP/#Introduction

based technology. The motivation lies in the fact that the current computational devices have failed to perform many basic tasks that biological systems have mastered, for instance, speech and image recognition. Thus, the next-generation computer design might borrow concepts from biological systems. On the hardware side, computer design might replace CMOS transistors using definite states (0 and 1) with quantum bits using superpositions of states for storing binary digits. These next-generation hardware architectures will require new programming models to utilize them.

References

1. Miner, D., Shook, A.: MapReduce Design Patterns: Building Effective Algorithms and Analytics for Hadoop and Other Systems, 1st edn. O'Reilly Media, Beijing (2012)
2. Hill, M.D., Marty, M.R.: Amdahl's Law in the Multicore Era. University of Wisconsinn, UW CS-TR-2007-1593, Madison (2007)
3. Gustafson, J.L.: Reevaluating Amdahl's law. Commun. ACM. 31(5), 532–533 (1988)
4. Berman, J.J.: Principles of Big Data. Elsevier/Morgan Kaufmann, Amsterdam (2013)
5. Consortium, "TIFF Specification, Revision 6.0," (1992)
6. Kumar, P., Alameda, J., Bajcsy, P., Folk, M., Markus, M.: Hydroinformatics: Data Integrative Approaches in Computation, Analysis, and Modeling. CRC Press LLC, Boca Raton (2006)
7. Bajcsy, P., Nguyen, P., Vandecreme, A., Brady, M.: Spatial computations over terabyte-sized images on hadoop platforms, In: 2014 IEEE International Conference on Big Data, (2014), pp. 816–824
8. Bajcsy, P., Vandecreme, A., Amelot, J., Nguyen, P., Chalfoun, J., Brady, M.: Terabyte-sized image computations on Hadoop cluster platforms. In: IEEE International Conference on Big Data (2013)
9. Blattner, T., Keyrouz, W., Chalfoun, J., Stivalet, B., Brady, M., Shujia, Z.: A hybrid CPU-GPU system for stitching large scale optical microscopy images. In: Parallel Processing (ICPP), 2014 43rd International Conference on, 2014, pp. 1–9
10. Altintas, I., Berkley, C., Jaeger, E., Jones, M., Ludascher, B., Mock, S.: Kepler: an extensible system for design and execution of scientific workflows. Sci. Stat. Database Manag. 2004 Proc. 16th Int. Conf. 1, 423–424 (2004)
11. Oinn, T., et al.: Taverna: a tool for the composition and enactment of bioinformatics workflows. Bioinformatics. 20(17), 3045–3054 (2004)
12. Talia, D.: Workflow Systems for Science: concepts and tools. ISRN Softw. Eng. 2013, 15 (2013)
13. Yang, Y., Xiang, P., Mantor, M., Zhou, H.: CPU-assisted GPGPU on fused CPU-GPU architectures. In: Proceedings – International Symposium on High-Performance Computer Architecture, (2012), pp. 103–114
14. Lee, J., Samadi, M., Park, Y., Mahlke, S.: Transparent CPU-GPU collaboration for data parallel kernels on heterogeneous systems. In: PACT '13 Proceedings of the 22nd International Conference on Parallel Architectures and Compilation Techniques, (2013), pp. 245–256

Supplementary Information

This chapter contains a summary of all web links related to software, test data, and deployed web systems.

Software and Documentation

(a) WIPP system download and instructions:

- Main WIPP website: https://isg.nist.gov/deepzoomweb/software/wipp.
- Docker container download and installation instructions are available from the section "WIPP deployment."
- User manual is available from the section "User guide."
- Step-by-step guidelines for several use cases from the section "Use cases."

(b) Source code on USNISTGOV Github:

- Location: USNISTGOV Github organization at URL – https://github.com/usnistgov
- Repository names:

  - Pyramid building algorithm (pyramidio): https://github.com/usnistgov/pyramidio
  - Stitching algorithm (MIST): https://github.com/usnistgov/MIST
  - Tracking algorithm (Lineage Mapper): https://github.com/usnistgov/Lineage-Mapper
  - Deep Zoom viewing and browser-based measurements (WebDeep ZoomToolkit): https://github.com/usnistgov/WebDeepZoomToolkit

- OpenSeadragon plugins to the framework available at http://openseadragon.github.io/#plugins

  – OpenSeadragonScalebar: https://github.com/usnistgov/OpenSeadragonScalebar
  – OpenSeadragonFiltering: https://github.com/usnistgov/OpenSeadragonFiltering

(c) Source code on isg.nist.gov web page:

- EGT algorithm: https://isg.nist.gov/deepzoomweb/resources/csmet/pages/EGT_segmentation/EGT_segmentation.html
- Accelerated stitching algorithm: plugin to ImageJ/Fiji – https://isg.nist.gov/deepzoomweb/resources/csmet/pages/image_stitching/image_stitching.html

(d) Project web page for WIPP

- Project description of the WIPP system: https://isg.nist.gov/deepzoomweb/resources/csmet/pages/web_image_pipeline/web_image_pipeline.html

Data for Testing Software Installation

(a) Small datasets for testing the WIPP installation

- Available from the main WIPP website at https://isg.nist.gov/deepzoomweb/software/wipp, section "Test datasets." The datasets will come classified in folders by job types. Each folder contains input data, expected results, and job configuration instructions.

(b) Datasets used in Chap. 2 (usage examples of the WIPP modules)

- These datasets are used for demonstrating the functionalities of the WIPP modules. They are available from the following URL (project web page for WIPP): https://isg.nist.gov/deepzoomweb/resources/csmet/pages/web_image_pipeline/web_image_pipeline.html.
- *Small dataset*: 5 × 5 Image Tile Dataset

    Cy5 Test Images (≈54 MB)   Phase Test Images (≈83 MB)

- *Large dataset*: 10 × 10 Image Tile Dataset

    Cy5 Test Images (≈119 MB)   Phase Test Images (≈195 MB)

Deployed Demonstrations on the Web

(a) Web Deep Zoom Toolkit deployment

   • This is the web user interface to Deep Zoom-based visualization of pyramids in WIPP with sample datasets.

      URL: https://isg.nist.gov/deepzoomweb/data

(b) Web statistical modeling deployment

   • This module is demonstrated with stem cell images.

      URL: https://isg.nist.gov/deepzoomweb/data/stemcellpluripotency

Printed in the United States
By Bookmasters